DRYLAND RESOURCES AND TECHNOLOGY

AIMS AND SCOPE

Drylands cover a vast area of earth. There is a growing awareness to have optimal exploitation of their resources and to ameliorate living conditions therein. This serial publication provides an international and multidisciplinary perspective of new ideas and technological developments relevant to drylands and their environments. Coverage widely includes topics such as natural resources, energy alternatives and applications; water management, resources development, desalination and usage (salinity, wastewater, etc.); land management and forestry; building and transportation; human and animal comfort; food and fodder; etc.

DRYLAND RESOURCES AND TECHNOLOGY

Editors

L.L. Somani
Professor of Soil Science
Rajasthan College of Agriculture
Udaipur - 313 001

K.N. Bansal
Associate Professor
Department of Agronomy
Rajasthan College of Agriculture
Udaipur - 313 001

R.A. Sharma
All India Coordinated Research Project for Dryland Agriculture
J.N. Krishi Vishwa Vidyalaya, College of Agriculture
Indore, 452 001 India

SCIENTIFIC®
PUBLISHERS

Published by

SCIENTIFIC PUBLISHERS (INDIA)
5 A, New Pali Road, P.O. Box 91
Jodhpur 342 001 (INDIA)
E-mail: info@scientificpub.com
Website : http://www.scientificpub.com

Print : 2022

ISSN: 053-5130

ISBN: 978-81-72330-82-8

eISBN: 978-93-88449-45-8

Reprinted from Dryland Resources and Technology
(Volume 8) 1994

Printed in India

Contents

Dryland Resources & Technology, 8 (1994)
Editors : L.L. Somani, K.N. Bansal & R.A. Sharma

SOIL MOISTURE USE BY CROP PLANTS

A. K. Sinha & Naveen Kalra
Water Technology Centre,
I.A.R.I., New Delhi – 110 012.

I INTRODUCTION

Crop production is generally determined by the prevailing environmental conditions i.e. by the existing complex of physical, chemical and biological factors. Water is one of the important components and is essential for plant survival and growth. Rain or irrigation water when stored in crop root zone flows from the soil through the roots and stems to leaves and then into the air for meeting the crop water demand which varies with soil-plant-atmosphere continuum characteristics of different agro-climatic zones. For realising optimum crop production it is necessary that soil water use by plants (consumptive use) matches the water demand (climatic evaporativity) since crop's response to water is related to the amount of water used by the crop.

Irrigation requirement and its scheduling depend considerably on consumptive use and other components of field water balance. The irrigation is becoming costlier day by day and it costs now approximately Rs. 35 to 40 thousand per hectare for development of canal irrigation. Since irrigation water needs to be applied to soil for use by plant, it is very necessary to have precise information on water use by different crop plants grown in different climatic condition with a view to work out proper consumptive use-irrigation crop yield inter-relationships. There have been world wide efforts on modelling the effect of water on crop yield to achieve the goal of crop production with maximum water use-efficiency.

II WATER IN SOIL-PLANT-ATMOSPHERE CONTINUUM (SPAC)

Water forms a dynamic common link in the soil-plant-atmospheric system and the use of soil water by plant occurs through a series of inter-related and interdependent processes. The water absorption is affected by rate of water loss and by the rate at which water moves from the soil to the root surface and the characteristics of SPAC system. Water either received from rain or irrigation is retained by soil according to its capacity as characterised by soil water storage capacity or field capacity (water held against-10 to -20 KPa matric potential in coarse textured soil and against -33 KPa in case of fine textured soil).

Due to evaporation from the soil surface, plant water withdrawal and redistribution within the soil profile, the moisture content in root-zone decreases and the suction of soil water with the solid matrix increases. With continuous depletion, a stage called permanent wilting point (moisture held against-1500 KPa) is reached when practically the water uptake by plant roots cease due to inability of roots to draw water against the interaction forces of soil matrix. There are reports suggesting water withdrawal by plant roots beyond permanent wilting point also. Water retained by soil between field capacity (upper limit) and permanent wilting point (lower limit) is termed as available water. The available water of soil varies with soil texture. It is as high as 0.23 cm/cm in case of fine textured soil. However, the total soil available water is a function of effective root-zone depth of crop in question, which is important for the operational scope of water use, critical soil moisture depletion and replenishment schedule.

Unlike soil, plant does not hold much water. A very insignificant amount compared to total water use by plants is held to maintain cell turgidity and metabolic activities. This water, of course, constitutes 80 to 90 percent of plant body. The main forces for its retention in plants are osmotic, turgor and matric (Tears and Peet, 1983). The plant water potential decreases as it advances away from root towards leaf and subsequently to atmosphere. Water moves from soil to atmosphere through plant due to potential gradient. This transport can be described by equation :

$$\text{Water flux} = \frac{\text{Difference in water potential between soil and atmosphere, } \Delta\Psi}{\text{Resistance, R}}$$

It is a convenient concept to use in describing water movement through the SPAC system because factors affecting water movement can be discussed in terms of their effects on either the driving force or the resistances. Under steady state condition, water transport in SPAC system can be written as :

$$Flow = \frac{\Psi soil - \Psi root\ surface}{R\ soil - root} = \frac{\Psi root\ surface - \Psi xylem}{R root - xylem}$$

$$= \frac{\Psi xylem - \Psi leaf}{R\ xylem - leaf} = \frac{\Psi leaf - \Psi air}{R\ leaf - air}$$

When there is ample amount of moisture present in the soil, water uptake by the plant is mainly governed by atmospheric evaporative demand (sink). As the moisture content in the root zone starts depleting, the source i.e. soil shows inability to meet the sink demand. Roots have to exert a higher pressure by lowering their potential appreciably to make the potential gradient at soil root interface higher but the soil-root interface resistance which depends upon the rooting density and water transmission characteristics of soil also increases. Plant adaptations can also be seen from increase of leaf water suction for drawing water from the source in tune with the atmospheric evaporative demand as well increase of stomatal resistance for water transport in situations when the demand is partially met from the soil. In summary, water flow through plants is controlled primarily in the vapor phase between the evaporating surface and the bulk air. Stomates are the principal regulators of water movement through plants. Decrease in water flow elsewhere especially in the soil and the roots, reduces transpiration indirectly by reducing leaf turgour enough to cause stomatal closure.

III FACTORS AFFECTING WATER USE BY CROP PLANTS

A. Root Characteristics

The important root characteristic that influences water uptake is depth wise distribution of effective roots. Root systems vary due to genetic constitution and micro- environmental conditions. Newman (1966) reported roots of cereals to be larger (200 to 4000 cm/cm^2) than nongramineous plants (15 -200 cm/cm^2). Under field conditions though roots absorb water from soils at different depths, root systems of crops are usually more concentrated in the surface layers which contain high proportion of available nutrients and are rewetted more frequently by irrigation or rain water. Nevertheless roots penetrate deep in the event of prolonged surface drying or the presence of shallow water table. On one hand when denser root systems deplete water rapidly, the deeper roots enable supply of larger volume of water for plant uptake than provided by surface layers. Variation in water supply is generally the major cause of differences in depthwise distribution of roots. According to Newman (1966), when water deficiency occurs in the upper soil layers, root growth is enhanced in the lower layers. However, the plants experiencing water stress in the early stages do not get an opportunity to develop a deep root system. Gajri and Prihar (1985) reported that higher root density of wheat in sandy loam soil increased leaf water potential suggesting increased uptake under a condition of even low available moisture. As mentioned earlier roots also offer resistance to water flow affecting uptake adversely. Favoured root growth at the expense of shoot growth under dryland condition as reported by various workers contending the surviving mechanism for drawing more amount of water from the profile. Deeper roots are reported to be more efficient in water uptake function being young, narrow and usually under wetter soil moisture regime than those closer to the surface (Taylor and Klepper, 1973). Water depletion effectiveness also decreases as the root becomes old (Allmaras *et al.*, 1975). Plants will preferentially take up water from the top soil layers if the water is amply available in those layers otherwise the atmospheric evaporativity demand will be met from the lower lying roots.

B. Shoot Characteristics

The water after entering root moves into stem and finally evaporates through leaves, where it encounters various resistances. In the stem the resistance is minimum whereas in leaves, the conductance of water through plasmodesmata is very low. Under a condition of nonlimiting water supply, the water losses through stomates are in accordance with photosynthesis and furthermore dictated by the prevailing environmental conditions. Diurnal variations in plant characters make the water movement in the plant system from soil highly dynamic in nature. Stomatal resistance quick tuning with the changes of the environmental condition is the major factor in this regard. Under the condition of moisture stress, water transport in the plant system is mainly dependent upon the availability of moisture in the root zone. Plant adaptation mechainsm such as reduction of leaf water potential resulting increase in resistances in the soil-plant system. Apart from this, the changes in canopy architecture in terms of leaf shape, size, orientation etc. also takes place under differential soil and environmental conditions.

C. Root-Zone Soil-Water Relations

Available soil water capacity in the root-zone is important for meeting consumptive use of crop. But with soil water depletion in the available range of soil- water, the supply of water to the root surface decreases to adversely affect the water uptake by roots. Availability of water to plants is significantly low as the depletion enters into the 3rd and 4th quarter of the available water range. This occurs because of decrease in hydraulic conductivity in unsaturated condition and the decrease is more marked in coarse textured soils. The soil water flux reaching to root surface is low at low moisture content and thus soil water use by plant is much less. Depending upon the soil, critical depletion levels are to be identified with optimum soil moisture use by crops for subsequently high production.

D. Atmospheric Evaporativity Demand

The atomospheric evaporativity is an index of rate at which evaporation of water from foliage and soil surface will occur. The diurnal variation in the atmospheric (sink) demand makes it dynamic. A number of research workers have computed climatic evaporative demand by using related weather parameters. Important one amongst these method is the reference evapotranspiration (ET_o) worked out by Modified Penman Method. The Reference Evapotranspiration (ET_o) represents the rate of evapotranspiration of an extended surface of 8 - 15 cm tall green grass cover, actively growing, completely shading the ground and not short of water and is given by the following relationship :

$$ET_o = c[W.Rn + (1-W).f(U).(e_a - e_d)]$$

where,

$(e_a - e_d)$ = vapor pressure deficit, mbar

f(U) = wind function

= 0.27 (1 + U/100) with U as wind speed km/day

Rn = net radiation

W = temperature and altitude dependent weighting factor

c = adjustment factor

The maximum evapotranspiration (ET_m) from a crop is related to ET_o by multiplying with crop factor, Kc the crop characteristics. Therefore, ET_m of any crop depends on values of ET_o which change with time in a cropping season. If the water is amply avaialble to the plant, the transport process will mainly depend upon the crop, its stage and existing weather conditions. Under inadequate moisture conditions, the transport process will be mainly governed by moisture situation alongwith environmental conditions.

IV. CROP YIELD RESPONSE TO SOIL MOISTURE USE

A. Adequate Soil Water

Adequate soil water refers to the moisture status in the soil that supplies water matching to the maximum evapotranspiration of crops. That is to say that actual evapotranspiration (ET_a) taking from crop is equal to the maximum evapotranspiration as determined by climatic parameters and crop characteristics. This condition remains till a certain depletion level of water in the available range and is dependent on crop, magnitude to ET_m and soil. In this condition there is no restriction to water availability to meet crop evapotranspirative requirement which is controlled by climate.

Very extensive research studies were conducted in India to work out permissible soil moisture depletion for irrigation scheduling of different crops. This was done through various indices like percent of available water, soil moisture tension, physiological stages, test plots and a few others. By and large it was found that permissible soil water depletion were 25% for vegetables and fodder crops, 50% for major cereals and pulses and 75% in case of most of oil seeds and some cereals like barley etc. In case of lowland rice it has been established that it was no more necessary to subject this crop to continuous standing water in field, but standing water of 5 ±2 cm may be depleted till 2 to 3 days below saturation before replenishment of water to the above level.

Doorenbos *et al.* (1986) divided crops in four groups according to soil water depletion (table 1) and suggested values of soil water depletion fraction (p) for crop groups and maximum evapotranspiration (ET_m) for ensuring $ET_a = ET_m$.

Table 1 : Crop Groups According to Soil Water Depletion

Group	Crops
1	Onion, peper, potato
2	Banana, cabbage, grape, pea, tomato
3	Alfalfa, bean, citrus, groundnut, pineapple, sun-flower, water melon, wheat
4	Cotton, maize, olive, safflower, sorghum, soybean, sugarbeet, sugarcane, tobacco.

Source : Yield Response to Water, FAO Irrigation and Drainage Paper No. 33

The consumptive use values of large number of crops have been worked out under conditions of deep, non-saline sandyloam soil of Delhi as presented in Table 2. The permissible level of soil water depletion has also been worked out and expressed in term of soil moisture tension. These values though worked out for sandy loam soil of Delhi are applicable in other soils and climatic conditions. The level of depletion may not coincide with the lower soil moisture limit for maintaining $ET_a = ET_m$, however, the values may be closer to it for ensuring maximum yield.

Table 2 : Soil moisture depletion level, water use and irrigation requirement of some major crops grown in deep non-saline sandy loam soil of Delhi

Crops	Moisture depletion level (-bars)	Percent ASM	Consumptive use (mm)	Economic yield (q/ha)*	Irrigation reqirement (mm)
Maize	0.65		412	43.7	125-206
Soybean	0.5 – 0.75		–		450
Sugarcane	0.33 – 0.70		980	710 (strippedcane)	667-810
Wheat (dwarf)	0.75 – 1.6	50	360	43	271
Wheat (tall)			315	33	
Potato	0.25 – 0.30		319	310	394-508
Cauliflower	0.5		200	283 (Coond)	
Onion	0.5 – 0.6		592	327 (bulb)	747
Tomato		40–50	406	400 (fresh fruit)	278
Green gram		50	–	–	140

Source : Basic Agronomic Research, Scientific basis of Agriculture : Seventyfive years of Service, IARI, 1980 : PP 152-55

Permissible soil moisture depletion for maximum water use and yield

In India considerable amount of research studies has been done on irrigation scheduling of crops in different agro-climatic regions. The extent of soil moisture depletion before irrigation for maximum crop yield has almost been established. Important criteria like percent available water depletion, soil moisture tension, irrigation depth, IW/Pan E ratio were used to express moisture stress/depletion before irrigation application. The values of these indices for some crops are given in table 3 wherein it is shown that in case of wheat permissible depletion of 50% of available moisture, soil moisture tension of 0.5 bar at 30 cm depth, IW/CPE = 0.75 to 1.0, and allowing moisture stress except at CRI, late tillering, late jointing and flowering stages gave maximum yield. However, in case of mustard depletion upto 75% of available moisture could be possible for maximum yield.

The consumptive water use is expected to be maximum when root- zone water is adequate and in such a situation maximum yields are obtained. The consumptive use values for different crops against permissible soil moisture level (soil moisture tension) are given in Table 2 as observed in sandy loam soil of Delhi. The values vary from 315 mm in tall wheat to 980 mm in sugarcane. The irrigation requirements differ with amount of rainfall. Through data presented in table 4 it is attempted to show the yield levels against irrigation applied following various criteria allowing soil moisture depletion for adequate consumptive use for maximum yields. The values of consumptive use were not reported. In case of lowland rice it has been established that it was no more needed to maintain continuous submergence necessitating application of irrigation water 2 to 3 times more than consumptive use. In such practice water requirement varies from 1520 to 2650 mm in the different parts of the country (Hukkeri and Pandey, 1977). The research conducted in various parts of the country has shown that maintaining soil moisture regime between standing water of 5 cm and depleting the same till 1 - 3 days following disappearance of water from soil surface before irrigation does not reduce yield significantly and maintains maximum consumptive use and saves irrigation water upto 20 to 65% as a result of reduction of percolation losses.

The presence of shallow water table with good water quality

Table 3: Criteria for scheduling irrigation to different crops.

Crop	Depletion of available water %	Physiological growth stage	Soil moisture tension (within 30 cm depth) bars	IW/PAN-E	Other Criteria
Wheat	25	C.R.I.	0.5	0.9	66-75% of
	50	late tillering late joining flowring		0.75	F.C. 122 cm
Barley	50	active tillering flag leaf flowering	–	0.8	–
Mustard	75	4-6 leaf stage			
		ranching/ pod formation	–	0.55	–
Berseem			0.3		
Cotton	25-50			0.8	
Sugarcane	30-50	–	0.7		
Gram	–	active bracketing /flower intiation		0.80	–
Maize	50	–	–	–	–
Groundnut	25-50	–	–	0.8-0.9	100 mm Cummulative PAN-E

Source : Soil-Water Relations by Khosla, B.K., Acharya, C.L. and Gupta, R.K. In. Review of Soil Research in India Part I 12th International Congress of Soil Science New Dehli 8-16 February, 1982.

was found to contribute considerably to meet the evapotranspiration demand of crops almost fully or partially (Pandey *et al.*, 1972). There was either no or very less irrigation required (Sinha and Singh 1982) (Table 5).

The crop yields of almost all crops under conditions of maximum consumptive use have been worked out against optimum irrigation schedules and recommendations have been reported by Rajput (1985).

Table 4 : Water use, Irrigation scheduling and yield of wheat in different parts of the country under adequate water supply conditions

Soil	Place	Irrig. schedule	No. of Irrig.	Total water applied (mm)	Yield (q/ha)
Alluvial (Sandy loam)	Ludhiana (Punjab)	Growth stages	6(7)	450	42.06
	Hissar (Haryana)	IW/CPE = 0.9	6(6)	360	38.56
	Delhi	IW/CPE = 0.9	5(6)	360	47.00
Desert (Loamy sand)	Jobner (Rajasthan)	IW/CPE = 1.05	13(4.5)	623	28.00
Vertisol	Rahuri	ET/PE = 2.05		403	40.2
	Navsari	–	7(7.5&10)	600	35.0

• The figures in parantheses is irrigation depth in cm

Table 5 : Optimum depth of water table for crops grown in sandy loam soil

Crops	Depth of water table (cm)•	Number of irrig. required
Sorghum	75	Nil
Cowpea	125	Nil
Maize	75-100	Nil
Bajra	125	Nil
Soybean	125-150	Nil
Paddy	25	Nil
Paddy	125	8
Wheat	50	Nil
Wheat	125	3
Peas	50-75	Nil
Toria	25	Nil
Mustard	125	1
Barley	150	2

A simplified approach has been adopted by dividing crops into four categories based on tolerance to soil moisture depletion as given in Table 1 and suggesting soil water depletion fraction (table 6) for each group of crops (Doorenbos *et al.*, 1986). No such grouping has been attempted in India and it may not be possible to adopt the above approach as such. In one study conducted in sandy loam soil of Delhi it was found that 25% more depletion than suggested by Doorenbos *et al.* (1986) could be possible in case of wheat.

Table 6 : Soil water depletion fraction (p) for crop groups and maximum evapotranspiraton (ET_m)

Crop	ET_m (mm/day)								
group	2	3	4	5	6	7	8	9	10
1	0.50	0.425	0.35	0.30	0.25	0.225	0.20	0.20	0.175
2	0.675	0.575	0.475	0.40	0.35	0.325	0.25	0.25	0.225
3	0.80	0.70	0.60	0.50	0.45	0.425	0.35	0.35	0.30
4	0.875	0.80	0.70	0.60	0.55	0.50	0.425	0.425	0.40

Source : Yield Response to Water, FAO Irrigation and Drainage Paper No. 33

B Limited Soil Water

The actual evapotranspiration (ET_a) matches the maximum evapotranspiration (ET_m) upto a certain level of soil water depletion and beyond this threshold value further depletion of water restricts the availability of water to roots and finally results into reduced water use by plants. Doorenbos *et al.* (1986) defined this condition of limited soil water through Available Soil Water Index (ASI) being less than 1 resulting into $ET_a < ET_m$. This situation occurs when there is insufficient rain and inadequate irrigation water availability such that application of irrigation is not possible before critical depletion levels. In this decreasing profile water content, water flux reaching to root surface for absorption decreases due to reduced hydraulic conductivity. The crop yields are adversely affected depending on the magnitudes of reduction in consumptive use as a result of decreased root water uptake. In such a situation of the inadequate water the approach has been the application of this limited water keeping in view inter- crop, inter-varietal and inter-stage responses to degree of stress for higher water use efficiency, adoption of measures for deeper rooting systems and other agronomic practices to channelise most of soil water to transpirational losses. The yield response to wheat with limited irrigation on different criteria is given in Table 7 for different soils. Further, decrease in irrigation reduces both water use and yield. Certain crops like mustard and gram respond better to wheat under limited water supply. For such a situation intercropping of wheat with mustard was found more remunerative in sandy loam soil of Delhi (Sinha *et al.*, 1985). In case of wheat it was found that stress tolerance was minimum during first sixty days compared to other growth periods and hence in case of one irrigation availability it was required to be applied here at IW/CPE = 0.75. Criticality of irrigation coincident with crown root initiation has been noted from stand points of root proliferation and nutrient availability (Chaudhary and Bhatnagar, 1986) and it opens the need for optimally planned deficit irrigation.

There are enough evidences of water extraction from deeper layers in the event of severe soil moisture stress (Prihar *et al.*, 1978, Sandhu *et al.*, 1978, Singh *et al.*, 1975). Infrequent irrigations were found to induce deeper rooting and greater depletion of subsoil water (Jalota *et al.*, 1979). Deeper storage of water resulted into greater soil water depletion and higher wheat yield than obtained from the same quantity of water stored in shallower soil depths seedzone for early establishment of root to enable penetration from dry surface layer to moist deeper layers for meeting water use by plants. The studies on mulching have clearly shown that it is possible to obtain good crop yield with very limited water.

Table 7 : Water Use. irrigation scheduling and yields of wheat under inadequate water supply situation

Soil	Place	Irrigation schedule	Irrigation No.*	Total water applied (mm)	Yield (q/ha)
Sandy loam	Ludhiana	Irrigation time table	3 (7)	310	41.3
	Hissar	IW/CPE = 0.6 or (CRI + Flowering.)	3 (6)	180	25.0
	Delhi	CRI + Flowering	3 (6)	180	41.3
Loamy sand	Jobner	IW/CPE = 0.6	9	420	19.6
Vertisol (black clay)	Sirguppa	IW/CPE = 0.6	–	–	39.3

* Figures in parenthesis show the depth of water at each irrigation

The yield response factor (Ky) (Ratio of relative yield decrease and relative evapotranspiration deficit) for different crops as advocated by Doorenbos *et al.* (1986) for different crops will enable to select appropriate crops for limited water conditions. They reported this factor for total growing period to vary from 1.25 in maize to 0.7 in groundnut suggesting that the latter is more efficient in water use and it may be preferred in the limited water condition.

V. MODELLING OF CROP WATER USE AND YIELD

Crops response to water supply has been extensively investigated giving consideration to soil hydraulic properties, crop characteristics as well as climatic conditions. But because of site specificity, the information has limited use under alien situations. In recent years, considerable progress has been made in quantitatively expressing crop performance by evapotranspiration functions that integrates many variables comprising the environment; thus making the findings amenable to be extended to diverse agro-climatic conditions. In most of the modelling exercises for crop water use, components of soil-plant-atmosphere continuum have been taken (Jensen *et al,* 1970; Ritchie, 1972, Angus *et al.*, 1980). Root water uptake from various layers have been predicted giving consideration to depthwise distribution of roots (Rasmussen and Hanks, 1978; Dejong and Cameron, 1979) but the activity of roots for water uptake has not been accounted for realistic estimate or root water withdrawl from various layers which calls for such consideration. Basic frame for development of root water uptake model is maintained by invoking the water balance equation :

$$\frac{\partial \theta}{\partial t} = \frac{\partial}{\partial z} \left(k(\theta) \frac{\partial h}{\partial z}\right) + A \ (z,t)$$

where θ is volumetric moisture content, h is soil water potential, t is time z is depth, K is hydraulic conductivity, $A\ (z,t)$ is root extraction term at depth z and time t. Integration of $A\ (z,t)$ over the root-zone depth at a particular time gives the transpiration by the crop. Nimah and Hanks (1973) predicted significant changes in root extraction, evapotranspiration and drainage by using

$$A\ (z,t) = [H\ (z,t) + 1.05\, z - h\ (z,t) - s\ (z,t)]\ \frac{RDF\ (z)\, k\ (\theta)}{\Delta x\ \Delta z}$$

where H is effective root water pressure head at the soil surface,

h is soil pressure head, s is salt potential, Δ x is distance between plant roots at the point in the soil (arbitrarily assumed to be one and *RDF (z)* is root distribution function in a depth integral Δz.

Molz and Remson (1971) considered sink term at each depth to be proportional to some effective roots distribution function as weight or length density and soil water diffusivity. Van Keulen (1975) expressed water uptake per soil layer as

$$\Delta q_r = T_m . L . \alpha . \beta$$

where T_m is maximum transpiration rate per unit of root length, L, α is temperature factor and β is reduction factor for water shortages.

Feddes *et al.* (1978) proposed a sink term by introducing a dimensionless sink term variable $\alpha (h)$ defined on the basis of soil pressure head, h as below

$$A(h) = \alpha (h) T_m / Zr$$

where T_m is the maximum transpiration rate and Zr is the rooting depth. This approach would be useful when roots are uniformly distributed over soil.

Jensen *et al.* (1970) defined actual evapotranspiration by multiplying potential evapotranspiration with the crop coefficient which depends on stage of crop as well as local climate and irrigation management conditions.

Hanks (1974) defined potential evaporation and transpiration under crop cover from the reference evapotranspiration on the basis of stage of the wheat growth. Actual evaporation was derived from potential value by using square root of time relationship since last wetting. Actual transpiration was deduced from potential value by incorporating the soil water availability as given below :

$$Ta = T_m \ if \ SWS / AW > / b$$

$$T_a = T_m \left(\frac{SWS}{AW x b}\right) \quad if \ SWS / AW < b$$

where SWS is soil water storage, AW is the maximum available water storage and b is the critical available water fraction, taken as 0.5 for wheat. Another approach of assigning potential transpiration to various root zone layers on the basis of root distribution factor and thereby judging the feasibility of respective layers on the basis of water availability and assigning the unsustainable load to the lower lying layers till the rootzone was covered has also been suggested.

Ritchie (1972) for sorghum suggested that in constant evaporation stage the actual evaporation from the soil surface in the crop canopy *(Es)* could be computed by :

$$Es = (0.92 + 0.4 \exp(-0.4\, LAI))\ \left(\frac{\Delta}{\Delta + \Gamma}\right) x\ Rn\ x\ \exp(-0.4\, LAI)$$

where LAI is leaf area index, *Rn* is net radiaiton, Γ is psychrometric constant and Δ is the slope of the saturated vapour pressure vs. temperature curve. In the falling rate stage, the square root of time relation could be employed. He further proposed actual transpiration as given below :

$$Ta = \frac{T_m}{1 + a \exp(-b\, AWF)}$$

For assessing yield on the basis of predicted water used by the crop, various production functions have been reported in literature. Choice of independent variables in this regard have been evapotranspiration, transpiration or applied water. In framing the production functions, both seasonal and inter-stage based approaches have been undertaken. In the seasonal based functions, both linear and non-linear (particularly quadratic) functions have been employed to plan irrigation schedule for a large command area but these functions lack in identifying the criticality of various stages towards water stress for yields (Dilman, 1931 and Tanner and Sinclair, 1983).

Interstage based production functios both multiplicative (Hanks, 1974; Jensen, 1968, Minhas *et al.*, 1984) and additive (Stewart *et al.*, 1977) have assessed the degree of senstivity to water stress for crop yields. Multiplicative models in this regard seem to be more realistic as the degree of stress in the earlier stage hampers the growth which even when multiplied with the subsequent luxuriant environment in the later stage might reduce the overall yield as in reality also the growth might not recover by the water availability in the later stages. This might not be reflected in the additive inter-stage response functions.

Stewart *et al.* (1977) proposed yield reduction factor (β) as ratio of yield reduction to evapotranspiration deficit as

$$\beta = \frac{(1 - Y/Y_m)}{(1 - ETa/ET_m)}$$

Doorenbos *et al.* (1986) utilized this for several crops to quantify the relationship between water and yield. Singh and Mann (1979) from a series of experiments on optimization of water use for crop production in arid region found linear relationship for yield and *ET* and linear to convexilinear yield-irrigation relationship. They also found the ratio of per cent yield reduction to per cent *ET* deficit (β) to be genetically reproducible character and arrived at 1.22 to be a generalised value of β for wheat and maize varieties. It was conlcuded by them that under limited water supply condition the deficits should be spread nearly evenly over the previous growth stages and the critical stage. Kalra (1987) reported yield reduction ratio value of 1.058 for wheat. Linear functions provided a good fit in case of seasonal evapotranspiration or transpiration based approach and no appreciable improvement was obtained by quadratic functions. Growth during mid season (62 - 112 day growth period) appeared most sensitive to water supply as judged from inter- stage based production functions. When canopy environment was included in the inter-stage transpiration based production functions, the quadratic form gave the best fit.

VI . CONCLUSION

Water use or consumptive use by crop plants is the key factor in crop production and a precise information about it is very essential for proper irrigation provision for optimum yields. fhe soil, crop and climatic characteristics influence the daily and seasonal evapotranspirational loss or consumptive use by influencing soil-plant-atmosphere continuum in a particular agro-climatic condition. The maximum water use (ET_m) by crop equals potential evapotranspiration (ET_o) upto a certain level of soil moisture depletion and any further depletion decrease actual consumptive use (ET_a). Crop growth and yield are adversely affected as the difference between ET_m and ET_a increases. The consumptive use and irrigation requirements for a number of crops have been found out by conducting experiments in different agro- climatic zones both for adequate and limited water supply conditions. However, yield response factors for these crops are required to be worked out for selection of crops based on water use efficiency. There is increasing interest in modelling of crop water use and yield on the lines as done in some other parts of the world so that crops can be grown for optimum yield vis-a-vis availability of water and other allied crop production factors.

REFERENCES

Angus, J.F, H.A. Nix, J.S. Russell and J.E. Kruizinga (1980) Water use, growth and yield of wheat in a sub-tropical environment. Aust. J. Agric. Res. 31 : 873-886.

Allmaras, R.R., W.W. Nelson and W.B. Voorhees (1975) Soybean and Corn rooting in South Western Minnesota. Soil Sci. Soc. Ann. Proc. 39 : 771-777.

Chaudhary, T.N. and V.K. Bhatnagar (1986) Wheat root distribution, water extraction pattern and grain yield as influenced by time and rate of irrigation. Agric. Water Management 2 : 115-124.

Dejong, R. and D.R. Cameron (1979) Computer simulation model for predicting soil water content. Soil Sci. 129 : 41-48.

Dilman, A.C. (1931) Water requirements of certain crops and weeds in the northern great plains. J. Agric. Res. 42 : 187-238.

Doorenbos, J. et al. (1986) Yield Response to Water. FAO Irrigation Drain. Pap. 33. Rome.

Feddes, R.A., P.J. Kowalik and H. Zaradny. (1978) Simulation of field water use and crop yield. Simulation Monographs. Pudoc. Wageningen.

Gajri P.R. and S.S. Prihar. (1985) Rooting, water use and yield relations in wheat on loamy sand and sandy loam soils. Field Crops Res. 12 : 115-132.

Hanks, R.J. (1974) Model for predicting plant yield as influenced by water use. Agron. J. 66 : 660-665.

Hukkeri, S.B. and S.L.Pandey. (1977) Estimating water requirements of crops. Chapter in Monograph "Water Requirement and Irrigation of crops" Monograph No. 4, W.T.C., IARI, New Delhi pp 91-162.

Jalota, S.K., S.S.Prihar, B.S. Sandhu and K.L. Khera, (1979) Water use and distribution of wheat as affected by pre-sowing and post-sowing irrigation. Agril. Water Manage 2 : 289-297.

Jensen, M.S. (1968) Water consumption by agricultural plants. In T.t. kazlowski (ed.) Water Deficits in Plant Growth, Vol. I. 1- 22. Academic Press, New York.

Jensen, M.E., D.C.N. Robband C.E. Frenzay (1970) Scheduling irrigations using climate-crop-soil data. J. Irrigation Drain. ASCE, 96 : 25-38.

Kalra Naveen, (1987) Evaluation of soil water status, plant growth and canopy environment in relation to variable water suppoy to wheat. Ph. D. thesis, IARI, New Delhi pp 103.

Minhas, B.S., K.S. Parikh and T.N. Srinivasan (1974) Towards the structure of a production function for wheat yield with dated inputs of irrigation water. Water Resources. Res. 10 : 383-393.

Molz, F.J. and I. Remson. (1970) Extraction term models for soil moisture use by transpiring plants. Water Resources. Res. 1346 : 1356.

Newman, E.I. (1966.) Relationship between root growth of linseed and soil water potential. New Phytol, 65 : 273-283.

Newman E.I. (1966) A method of estimating the total length of root in a sample. J. Appl. Ecol 3 :139-145.

Nimah, M.N. and R.J. Hanks. (1973) Model for estimating soil water, plant and atmospheric inter-relations. Soil Sci. Soc. Am. proc. 37 : 522-527.

Pandey, S.L.,A.K. Sinha and M. Pal (1972) Water table flucutations and its interpretation. Symp on Water logging causes and measures for its prevention 45th Annual Convention of CBIP, Publication No. 1180 : 55-80.

Rajput, R.K. (1985) Research on Water Management. Progress report of Coordinated Project for Research on Water Management, Bhartiya Krishi Annusandhan Bhavan, New Delhi.

Rasmussen, V.P. and R.J. Hanks. (1978) Spring Wheat yield model for limited moisture conditions. Agron. J. 70 : 940-944.

Ritchie, J.T. (1972) Model for predicting evaporation from a row crop with incomplete cover. Water Resource Res. 8 . 1204 1213

Sandhu, B.S. and S.S. Prihar (1983) Irrigation scheduling to crops in Punjab, Bulletin Department of Soils, P.A.U., Ludhiana, pp 81.

Sandhu, B.S., S.S. Prihar, K.L. Khera and K.S. Sandhu, (1978) Scheduling irrigation to chickpea. Indian J. Agric. Science 48 : 486-492.

Singh N.T. and G.S. Brar, (1979) Effect of soil moisture regime on xylem water potential of three wheat varieties. Indian J. Agric. Sci. 49 : 266-271.

Singh R.Y. Sinha, S.S. Prihar and P. Singh (1975) Effect of N fertilization on yield and water use efficiency of dryland winter wheat as affected by stored water and rainfall. Agron. J. 67 : 599-603.

Singh S.D. and H.S. Mann. (1979) Optimization of Water Use and Crop Production in an Arid Region. Res. Bulletin 1, CAZRI. Jodhpur : pp 88.

Sinha A.K. and N.P. Singh (1982) Salient Research Achivements of Coordinated Project on Water Management (1969-81). W.T.C. IARI, New Delhi pp 19.

Sinha, A.K., K.K. Nathan and A.K. Singh (1985) Radiation climate and water use studies in inter-cropping systems. J. Nuclear Agric. Biol 14 : 64-69.

Stewart, J.I., R.J. Hanks, R.E. Danielson, E.B. Jackson, R.M. Hagan, W.O. Pruitt, J.P. Riley and W.T. Franklin (1977) Utah Water Res. Lab. Report PRWG 151-1, Utah State Univ. Logan.

Tanner, C.B. and T.R. Sinclair, (1983) Efficient water use in crop production research. In. H.M. Taylor et al. (eds) Limitation to Efficient Water Use in Crop Production, A.S.A.E. Monograph.

Taylor, H.M. and B. Klepper. (1973) Rooting density and water extraction pattern for corn. Agron. J. 65 : 965-968.

Teare I.D. and M.M. Peet. (1982) Crop Water Relations. John Wiley and Sons, New York, pp 547.

Van Keulen, H. (1975) Simulation of water use and herbage growth in arid regions. Simulation Monographs. Pudoc. Wageningen.

Dryland Resources & Technology, 8 (1994)
Editors : L.L. Somani, K.N. Bansal & R.A. Sharma

EFFECT OF WATER DEFICIT ON SYMBIOTIC NITROGEN FIXATION IN LEGUMES

K. Swaraj & N.R. Bishnoi,
Department of Botany,
HAU, Hisar - 125 004

I INTRODUCTION

Legumes in India are generally cultivated under rainfed conditions in the semiarid tropics where the plants are afflicted with scarce water availability and high subsoil temperatures reaching 40°C during summer. Bengal gram (7.1 million hectares) and red gram (2.1 million hectares) which contribute more than half to the annual pulse production in India are invariably grown in areas faced with prolonged and frequent periods of drought where application of nitrogenous fertilizers is uneconomical and the plants have to meet their nitrogen requirement through symbiotic nitrogen fixation. Symbiotic nitrogen fixation in legumes is determined by inherent characteristics of both the host plant and the rhizobial strain. Genetic variation in the two partners leads to development of symbiotic association with a variable functional efficiency. The functioning of the symbiotic system is affected by even mild stress conditions which have no adverse effect on growth of plants dependent on combined nitrogen. All the stages of symbiosis including pre-infection stages are adversely affected by periodic environmental stress of varying duration and intensity during the growth period of the plant.

In the past one decade lot of scientific information has been generated in connection with the physiology of symbiotic system in legumes under stress condtions. The work has earlier been reviewed by Sprent (1981) and Swaraj (1987). However, in the meanwhile more information has appeared and in view of the importance of the subject the topic needs reviewing again.

II EFFECT OF WATER DEFICIT ON NODULATION

A complex physical interaction between the two symbiotic partners i.e. the host legume and a compatible rhizobial strain is a pre- requisite for the nodule initiation. First rhizobia must move to the vicinity of the host root system and this movement is considerably restricted under high soil moisture tensions (Hamdi, 1971; Bushby, 1982) which might ultimately reflect itself in reduced nodule number. In *Trifolium subterraneum* a decrease in soil water potential from - 36 to - 360 kPa significantly decreased the number of infection threads formed and inhibited nodulation completely (Warrall and Roughley, 1976). The optimum moisture level for symbiosis in lucerne and soybean (Fred *et al.*, 1932) and *Acacia* (Habish, 1970) was found to be around 62-75 per cent of the water holding capacity of the soil and any deviation from this optimum level adversely affected nodulation.

Wilson (1931) found in a glass house study with *Phaseolus vulgaris* that soil dehydration to various soil water contents before watering could result in a decrease in nodule number.

Venkateswarlu (1985) reported that increasing moisture stress by decreasing soil water potential to less than -1.5 MPa did not effect the nodule in mungbean, mothbean and clusterbean significantly but under maximum moisture stress level a 65 per-cent decrease in nodule fresh weight occurred due to dehydration and cessation of meristematic activity. He also reported that in these crops under field conditions stress at the seedling stage delayed nodule initiation by 3-4 days and reduced the nodule number too. At flowering, reduction in nodule fresh weight was less in clusterbean (54%) compared to mungbean (86%) and mothbean (62%) for an equivalent degree of stress. Patterson *et al.* (1979) studied the response of growth chamber grown soybean plants to a 4 day drought. Nodule growth in terms of dry weight seemed to be suspended during drought but nodule growth was resumed at control rates on rewatering. No shedding of nodules occurred as a result of short term drought. The studies carried out on field grown soybeans for three years showed that mild drought had no effect on nodule number or dry weight. Sprent (1976) Kulkarni *et al.* (1988) and Sheoran *et al.* (1988) also reported a decrease in nodule number and size under moisture stress conditions. But it was not known whether reduction in nodule number resulted from fewer rhizobia available for infection or from effects on the infection process.

Indeterminate type of nodules like those of *Vicia, Pisum* and *Trifolium* which are endowed with prolonged meristematic activity have been reported to be relatively more resistant to drought (Sprent, 1972; Engin and Sprent, 1973) than determinate spherical nodules with limited meristematic activity like those of soybean and cowpea (Sprent, 1976). Indeterminate nodules could recover from relatively severe moisture stress because of restoration of the meristematic activity on replenishment of the soil moisture. Determinate nodules are irreversibly damaged by equivalent levels of stress and ultimately shed. The plant has to produce a new crop of nodules. Renewed root nodule formation could occur in different soil layer which might be less affected by subsequent stress. The formation of new nodules could be a serious drain on photosynthates (Sprent, 1981) which might have otherwise been diverted to plant growth.

Studies on detached nodules showed that desiccation resulted first in the breakdown of plasmodesmata interconnecting plant cells followed by shrinkage and ultimately disintegration of cytoplasm of cortical cells surrounding the bacteroidal tissue

(Sprent, 1971, 1972). Collapsed cortical cells might block the connection of the intercellular channels serving as gas exchange ports between the central bacteroidal tissue and the rhizosphere and thus creating a physical barrier to O_2 diffusion inside the nodules. This hypothesis was supported by observations of Pankhurst and Sprent (1975) who showed that the rate of N_2- fixation by the bacteroids isolated from moderately stressed and control nodules was same. In addition, inhibition of N_2-fixation in moderately stressed nodules could recover completely and in severely stressed nodules it could recover partially under high pO_2. Guerin *et al.* (1990) on the other hand reported that the bacteroids isolated from stressed nodules were more sensitive to O_2. They could observe only partial recovery of C_2H_2 reduction capacity of mildly water stressed nodules (-1.2 MPa) under high pO_2.

Prolonged periods of water deficit accelerated the rate of nodule senescence (Sprent *et al.*, 1983; Sutton, 1983; Ahlawat, 1990). Shedding of determinate and indeterminate nodules of both pasture and grain legumes might result from severe stress induced senescence.

Electron microscopic studies carried out by the senior author in Timiriazev Institute of Plant Physiology, Moscow revealed that under moisture stress peribacteroid membrane was the first to show a breakdown followed by disintegration of the host cell cytoplasm. The bacteroids did not show disintegration even under severe stress conditions. However, many times, they were literally loaded with poly-B-hydroxybutyrate crystals.

Pronlonged periods of moisture stress resulted in a decreased nodule weight and its dichotomous branching in chickpea (Ahlawat, 1990). Thickness of cortical tissue decreased and the endodermis became discontinuous under moisture stress conditions.

A. DROUGHT EFFECTS ON NITROGENASE ACTIVITY AND RELATIVE NITROGEN FIXING EFFICIENCY

Drought has been shown to affect N_2-fixation adversely in case of a number of legumes such as *Glycine max* L. (Sprent, 1972; Huang *et al.*, 1975a, b; Albrecht *et al.*, 1984; Lawrence and John, 1988; Deveries *et al.*, 1989) *Lupinus arboreus, Medicago sativa* L. (Engin and Sprent, 1973; Aparcio-Tejo and Sanchez-Diaz, 1982), *Cajanus cajan* Mill sp. (Sheoran *et al.*, 1981; Nandwal, 1989), *Arachis hypogaea* L.

Table 17 : Effect of gypsum application on soil chemical properties.

Gypsum (% GR of Soil)	pH (1 : 2) B. S.*	pH (1 : 2) A. H.**	EC (1 : 2) dSm^{-1} B. S.*	EC (1 : 2) dSm^{-1} A. H.**	ESP B. S.*	ESP A. H.**
0	9.59	9.59	0.23	0.31	42.0	44.5
25	8.88	8.90	0.34	0.34	25.0	27.0
50	8.10	8.30	0.50	0.61	12.0	12.5
75	7.93	8.00	0.84	0.98	10.0	10.5
100	7.78	7.90	1.04	1.20	9.0	9.0

*Before sowing **After harvest

The efficiency of gypsum can be improved when applied in proper time and amount with suitable method. Placing gypsum fragments or rock gypsum in enlarged section of water courses was found useful for amelioration of water having a high Na:Ca ratio (Pal and Poonia, 1979). It was found that application of gypsum at each irrigation was better than applying gypsum in one dose under a fodder-maize rotation (Bajwa *et al.*, 1981, 1983). However, Bajwa and Josan (1989) reported that though, there was no effect of different rates and methods of gypsum application on wheat yield, but the beneficial effect of gypsum applied to rice in one dose was less as compared to application with each irrigation. Small repeated application of gypsum to the soil may be effective in surface infiltration problems (Ayers and Westcot, 1985).

Gypsum to supply 2.5 and 5.0 meq Ca L^{-1} of sodic irrigation water for wheat and rice crops, respectively, was sufficient to maintain high yields (Bajwa and Josan, 1989). Manchanda *et al.* (1989) studied the effectiveness of soil application of gypsum at five rates (0, 25, 50, 75 and 100% GR of soil) in terms of performance of wheat, bajra and guar on two sites having sodic irrigation water and soil and concluded that these crops could be grown sucessfully provided gypsum was added 100% GR of the soil plus that required to neutralise entire RSC on cumulative basis, before the onset of monsoon. Yadav *et al.* (1991a) observed that addition of gypsum 50% GR of soil irrigated with sodic water (EC = 2 dSm^{-1}, RSC = 12 me L^{-1}) was appropriate for growing most of the *kharif* crops (Table 18). They further added that gypsum application before monsoon was the best time followed by before presowing irrigation of *rabi* crops and with each irrigation.

Table 18 : Effect of gypsum application on the grain yield of crops under sodic water irrigation.

Gypsum (% GR of soil)	Grain yield (q ha^{-1})					
	Pearl-millet	Moong bean	Urd bean	Cow-pea	Pigeon-pea	Cluster bean
Control	18.75	0.96	1.32	3.23	0.11	10.01
25	26.95	1.99	2.38	6.15	0.24	16.17
50	29.25	1.87	2.61	8.84	0.35	17.32
100	22.90	1.49	2.19	7.50	0.19	23.10
CD (= .05)	7.70	0.69	1.04	3.81	0.22	3.48

However, Sharma *et al.* (1988) found that medicinal plant (*Hysoscyamus spp.*) gave highest herbage yield and economic returns when gypsum was applied 75% GR of soil. Bajwa and Josan (1989) studied the effect of gypsum application on wheat and rice yields in relation to two sodic waters (W_1 & W_2) having respectively, EC of 0.85 and 0.95 dSm^{-1}, SAR of 8.4 and 13.5, and RSC of 6.8 and 10.3 meq L^{-1}. They observed that average responses of both rice and wheat to gypsum were greater under W_2, and it increased over the years. Under W_1 treatment, yields of both crops with gypsum applied either cumulatively or at each irrigation, were equal. Under W_2 the effect was similar in wheat only. But in case of rice, gypsum applied at each irrigation with W_2 was better than that applied in one dose. Lower rice yields under one single application could be due to an appreciable increase in Na saturation of the soil with sodic irrigation water without gypsum during the growing season. Whereas, in case of wheat the depth of irrigation water used during the growing season was much smaller, and increase in Na saturation of the soil was not sufficient to adversely affect the yields.

Keren and O'Connor (1982) reported that the rate of gypsum dissolution is influenced by both gypsum content (for a given fragement size) and soil water velocity. However, Hira and Singh (1980) found that amount of water required to dissolve the added gypsum and to reclaim a sodic soil did not increase with the amount of gypsum added, provided it equalled the gypsum requirement. The increase in the soil water velocity decreased the contact time between gypsum particles and moving water leading to corresponding decrease in value of dissolution rate constant (Kemper *et al.*, 1975). About 35 cm water was required for dissolution of gypsum applied to an air dry sodic clayey soil in the quantities less than or equal to gypsum requirement (Gupta and Ranade, 1987). While Hira and Singh (1980) reported that dissolution of gypsum depended on particle surface area. The water requirement increased from 2.8 cm to 15.9 cm as gypsum particle size increased from 0.1 mm to 0.5-2.0 mm.

Sharma *et al.* (1988) observed that increasing rate of gypsum had greater beneficial effect on plant population, plant height, number of branches per plant, dry matter yield per plant in case of *Hysoscyamus albus*. Yadav *et al.* (1990b) reported that the germination, nodulation and yield of chickpea improved with increasing rates of gypsum upto 50% GR of soil (Table 19).

Use of gypsum along with improved package of practice has shown that good crop of rice and wheat can be grown on highly deteriorated alkali soils (Abrol and Bhumbla, 1979). Sawhney and Baddesha (1989) conducted a long term experiment on farmers field on highly saline sodic soil and argued that gypsum application increased the grain yield of wheat from 25.5 to 43.8 and of rice from 34.2 to 68.6 q ha^{-1} during three year of reclamation. Yadav *et al.* (1991) conducted field experiments on a sodic loamy sand soil (pH 9.6-9.7) with gypsum application (complete neutralization of RSC cumulatively) for two consecutive years at Bawal (Haryana) by growing major *rabi* and *kharif* crops. They found that addition of gypsum enhanced the grain yield of mustard to the maximum (17.8 q ha^{-1}). Among *kharif* pulses clusterbean and cowpea gave good yield but pigeopen, moongbean and urd bean yielded poorly (Table 20).

Sharma and Yadav (1986) observed that application of gypsum markedly increased the grain yield, straw yield and uptake of Fe, Mn and P by rice plant over control. However, the grain yield, harvest index and uptake of nutrients were found to be highest when gypsum was applied at the rate of 7.5 g kg^{-1} soil. They further argued that at higher doses of gypsum (15g kg^{-1} soil) the higher concentration of Ca probably induced the imbalance in Fe, Mn and P uptake and depressed the grain yield. (Table 21). Similar trend in rice yield was earlier reported by Shahi *et al.* (1978), Abrol and Bhumbla (1979) and Singh *et al.* (1980).

Table 19 : Effect of gypsum and phosphorus application on growth attributes of chickpea

Gypsum (% GR of Soil)	Levels of Phosphorus (Kg P_2O_5 ha^{-1}) 0	20	40	60	Mean
	Per cent germination				
0	66	63	66	54	62.0
25	76	79	88	68	77.8
50	93	93	96	78	90.0
75	80	85	82	66	78.3
100	77	80	78	55	72.0
Mean	78.4	80.0	81.6	64.0	
	Nodule fresh weight (g pot^{-1})				
0	5.5	8.5	8.5	8.0	7.6
25	7.5	14.7	14.5	14.7	12.8
50	17.2	18.5	19.5	18.5	18.4
75	22.0	24.5	28.5	29.0	26.0
100	17.0	19.2	18.0	19.5	18.3
Mean	13.8	17.0	17.8	17.9	
	Grain yield (g pot^{-1})				
0	6.3	9.2	10.2	10.2	8.9
25	8.4	9.8	10.6	10.8	9.9
50	9.3	10.9	11.6	11.8	10.8
75	9.6	10.9	11.7	12.2	11.1
100	8.2	9.3	10.9	12.0	10.1
Mean	8.4	10.0	11.0	11.4	

CD (= .05)	% germination	Nodule weight	Grain yield
Gypsum	12.0	5.2	0.8
Phosphorus	9.0	3.1	0.6

Table 20 : Productivity of oil seeds and pulses under sodic waters with and without gypsum.

Treatment	Grain yield (g. ha^{-1})							
	Raya	Chick-pea	Field-pea	Cluster bean	Cow-pea	Moong bean	Urd bean	Pigeon pea
No gypsum	5.3	0.1	1.4	10.0	3.2	1.0	1.3	0.1
Gypsum	23.2	6.3	14.6	23.1	8.8	1.9	2.6	0.4

Source : Yadav et al. 1991.

Table 21 : Effect of gypsum on grain yield, straw yield, harvest index and Fe/Mn and P/Fe ratios in rice grain.

Gypsum levels (g kg^{-1} soil)	Grain yield (g pot^{-1})	Straw yield (g pot^{-1})	Harvest index	Fe/Mn	P/Fe
0	19.3	39.8	0.30	5.84	5.65
7.5	95.6	82.9	0.51	6.67	5.42
15	70.1	94.1	0.43	6.26	6.12

(b) Pyrites

Use of pyrite has given encouraging results in the reclamation of sodic soil (Mehta and Abrol, 1976; Venkatakrishnan and Abrol, 1981; Swarup, 1985). For sodic soils growing sugarcane, application of pyrites (4 mg/ha^{-1}) has given encouraging results (Rai *et al.*, 1980). Garg (1988) reported higher yield of top, root and sugar in the beets grown on soils amended with pyrites. Composition of minerals was higher in tops (except Mg) and roots (except Na) in pyrite amended soils. Alam *et al.* (1986) observed that there was an appreciable decrease in pH, EC, ESP, CO_3, HCO_3, RSC and increase in Ca + Mg and SO_4 in saturation extract and available soil P due to application of pyrites which also increased the yield of rice and wheat at different levels of pyrites. They further added that beyond a certain level of pyrites application, the trend in soil physical improvement and increase in yield of crop was reverse. Verma and Gupta (1985) studied the relative effectiveness of pyrites in reclaiming sodic clay soil and reported that decrease in soil pH with time was more in pyrite treated soil as compared with gypsum applied one probably due to the formation of sulphuric acid (Mitra and Shanker, 1957).

(c) Farm Yard Manure

Soil rich in sodium are characterised by poor physical condition resulting in deflocculation, dispersion and Na induced swelling (Gupta and Verma, 1983). Kesvan and Gupta (1986) suggested that though organic matter present in soil could not prevent sodification (rather enhanced the process), its application reduced the soil ESP in already sodified soils. Poonia *et al.* (1984) reported from a laboratory study that sodification at a given SAR was less for soils containing high organic matter. Though organic matter addition reduces the preference of soil colloid for Na (Gupta *et al.*, 1984), yet it increases clay dispersion under sodic conditions (Poonia *et al.*, 1982). Puntamkar *et al.* (1972) and Dhankhar *et al.* (1990) observed the beneficial effect of FYM application on the utilization of sodic waters in wheat while Manchanda *et al.* (1985) found no such effect in the crop supplied with recommended dose of fertilizer.

(d) Green manuring with dhaincha

Dhaincha has generally been recommended as an ideal salt tolerant crop for manuring the sodic soils (Agarwal *et al.*, 1950; Uppal, 1955). The combined effect of gypsum and dhaincha in reclaiming such soils was reported by Agarwal and Mehrotra (1953). The pH of dhaincha leaves is highly acidic (4.0) hence it effectively reduces the pH of sodic soils. Its extensive root system proliferates deep in the soil and helps in better aeration of plants and leaching of salts. In addition to this, it is moderately tolerant to water logging which is a common feature of soils irrigated with sodic waters. Yadav and Aggarwal (1961) reported that green manuring with dhaincha showed an additional yield increase of 13.91% in paddy and 21.45% in berseem fodder in subsequent *rabi* over the application of gypsum alone. (Table 22). The content of nitrogen, organic carbon and permeability of soil increased appreciably in the treatment where gypsum was supplemented with dhaincha.

Table 22 : Effect of green manuring with dhaincha on paddy and berseem yields

	Paddy				Berseem	
	Grain		Straw		Fodder yield	
	Without dhaincha	With dhaincha	Without dhaincha	With dhaincha	Without dhaincha	With dhaincha
First Year						
Yield (mds acre^{-1})	13.71	15.62	30.51	32.54	69.57	84.49
Percent increase over gypsum	100.00	113.91	100.00	106.64	100.00	121.45
Second Year						
Yield (mds acre^{-1})	20.54	26.74	83.52	94.42	76.46	98.01
Percent increase over gypsum	100.00	130.19	100.00	113.11	100.00	128.15

(e) Inorganic fertilizers as amendments

Sodic soils are known for their poor physical conditions, they are reported to be low in organic matter and available N and contain high amounts of extractable P and K (Chhabra, 1985). Extractable P is also reported to decrease in surface soil, through leaching, during the process of reclamation. Use of fertilizers particularly the phosphatic ones, is known to improve soil structure by promoting aggregation in normal soils (Thein, 1976). While inorganic nitrogenous fertilizers have shown depressing effect on soil physical properties phosphorus application in conjuction with N plays an important role in checking these deleterious effects (Biswas *et al.*, 1979). It has been observed that application of fertilizers lowered the soil pH and significantly improved physical properties of gypsum amended soil (Chawla and Chhabra, 1991) and that P alongwith N further increased infiltration rate, saturated hydraulic conductivity and percentage of water stable aggregates, whereas bulk density, water dispersible silt plus clay content and modulus of rupture decreased. There was no significant effect of K application on physical properties of soil (Table 23).

Table 23 : Soil properties as affected by various fertilizer treatments.

Treatment	pH	EC (dSm^{-1})	Avail. P ($kg\ ha^{-1}$)	Avail.K ($kg\ ha^{-1}$)	O.C. (%)	B.D. ($Mg\ m^{-1}$)	Sat.K ($mm\ d^{-1}$)
			0–7.5 cm				
Control	8.7	0.52	7.5	382	0.56	1.38	0.40
$R_N W_N$	8.5	0.41	3.8	297	0.59	1.36	2.09
$R_N W_{NP}$	8.4	0.44	14.7	281	0.60	1.35	3.05
$R_N W_{NPK}$	8.5	0.49	17.5	312	0.52	1.33	3.59
$R_{NP} W_{NP}$	8.4	0.41	29.0	268	0.56	1.31	6.22
$R_{NPK} W_{NPK}$	8.4	0.42	35.3	339	0.56	1.31	6.11
CD = 0.05	0.1	NS	3.4	32	NS	0.04	2.42
			7.5–15 cm				
Control	8.9	0.54	8.5	386	0.44	1.56	0.00
$R_N W_N$	8.6	0.43	3.5	288	0.43	1.56	0.05
$R_N W_{NP}$	8.6	0.44	10.5	280	0.44	1.56	0.15
$R_N W_{NPK}$	8.6	0.49	11.1	295	0.48	1.55	0.25
$R_{NP} W_{NP}$	8.5	0.41	20.1	271	0.47	1.54	0.40
$R_{NPK} W_{NPK}$	8.5	0.46	22.7	317	0.41	1.54	0.41
CD .05	0.2	NS	3.3	37	NS	NS	0.12

R =Rice; W =Wheat; N, P, K =Full dose of N, P or K for respective crops.

(f) Relative efficiency of amendments

Puntamkar *et al.* (1972) studied the effect of gypsum and manure on the growth of wheat irrigated with bicarbonate rich water and reported that both the amendments increased the yield of wheat separately and highest yield of wheat was recorded with 20 tons of gypsum alone. Combined gypsum and manure were less valuable (Table 24). Similar results were reported by Yadav (1989) in bajra.

Table 24 : Effect of gypsum and manure on the yield ($t\ ha^{-1}$) of wheat.

Manure ($t\ ha^{-1}$)	Gypsum ($t\ ha^{-1}$) 0	5	10	20
0	3.00	3.66	4.20	5.26
15	3.40	4.06	4.80	5.13
30	3.80	4.40	4.73	5.06

Yadav (1991) observed that the response of gypsum increased with the corresponding increase in RSC content of the irrigation water reverse was the case with FYM (Table 25).

Table 25 : Effect of gypsum and FYM application on the grain yield of wheat in relaiton to irrigation with sodic water of variable RSC.

Treatment	RSC levels (me L^{-1})			
	2.8	12.0	16.0	Mean
Control	42.2	33.3	19.4	24.1
Gypsum	–	43.6	30.5	33.0
FYM	46.8	35.9	20.7	31.2
Mean	44.5	37.6	23.5	
CD (=.05) RSC=4.4:	Amendment = 4.4:	RSC x Amendment = 6.3		

Source : Yadav (1991).

In pots, Patel and Singh (1991) studied the effect of application of gypsum, pressmud and pyrites on soil properties and yield of rice and found the residual effect on wheat. They reported that gypsum was more effective in removing leachate, Na and Ca + Mg whereas the cumulative removal of CO_3 +HCO_3 was higher with pressmud treatments than that of gypsum and pyrites under percolated condition. They further added that gypsum was more effective in reducing the pH, ESP and increasing exchangeable Ca and Mg in the soil than pressmud and pyrites. The highest yield of rice was recorded with pressmud followed by gypsum and pyrites (Table 26). Pyrite was found to be as effective as one third of gypsum when applied on total sulphur content basis.

In a similar study, Swarup (1985) compared the performance of some amendments in a highly sodic soil (ESP 94, pH 10.6) and rated their effectiveness as gypsum > pyrites > FYM > rice husk > Control. It was observed that the degree of improvement was further enhanced with the duration of flooding and crop growth. They further observed that incorporation of FYM resulted in marked improvement of Fe, and Mn concentration of crop over control, gypsum and pyrite and, this was followed by rice husk (Table 27). The sodium content of grain and straw decreased with the application of amendments. He found that gypsum is more effective in reclaiming the sodic soil by lowering the ESP and increasing exchangeable Ca + Mg than pyrites, FYM and rice husk.

Table 26 : Effect of amendments on grain and straw yield (g pot^{-1}) of rice and wheat

Amendment level (% of GR)	Rice		Wheat	
	Grain	Straw	Grain	Straw
Control	6.3	15.7	0.54	1.91
Gypsum (25)	22.1	30.1	5.43	9.37
Gypsum (50)	30.0	39.6	7.14	14.22
Pressmud (25)	22.7	33.2	5.13	9.07
Pressmud (50)	33.0	43.7	6.10	12.51
Pressmud (75)	34.7	45.1	7.24	14.60
Pyrites (25)	11.5	23.3	2.92	5.13
Pyrites (50)	19.6	27.5	4.40	7.83
Pyrites (75)	23.4	31.2	5.27	9.87
CD (=.05)	1.48	1.75	0.52	0.78

Table 27 : Chemical composition of rice grain.

Treatment	P(%)	K(%)	Na(%)	Ca(%)	Mg(%)	Zn(ppm)	Cu (ppm)
Control	0.29	0.38	0.85	0.08	0.07	15	9
Gypsum	0.34	0.39	0.09	0.14	0.14	30	12
Pyrites	0.35	0.38	0.10	0.13	0.13	28	12
FYM	0.32	0.40	0.11	0.12	0.11	27	10
Rice husk	0.30	0.39	0.14	0.11	0.10	24	11
CD (=.05)	0.02	NS	0.02	0.01	0.01	3	1

Singh *et al.* (1986) studied the effect of pyrites and sulphitation pressmud in calcareous saline-sodic soils and found optimum dose of 4 t of pyrites or 20 t of sulphitation pressmud per hectare which increased the yield of sugarcane to the tune of 19.45 and 20.52 t/ha over control giving economic gain of about Rs. 3,467 and 3,601, respectively. There was an increase of 0.5 to 1.5 units of sugar per cent in cane and 1.5 to 3.0 t/ha estimated commercial cane sugar due to application of these amendments. Verma *et al.* (1985) studied the relative effectiveness of some chemical amendments in reclaiming sodic vertisols and found that in reducing ESP, their efficiency varied as $Al_2(SO_4)_3$ > gypsum > H_2SO_4 > Pyrites > $FeSO_4$ > Control. The lesser efficiency of H_2SO_4 as compared to $Al_2(SO_4)_3$ and gypsum is due to rapid reaction of acid with soluble Na_2CO_3 and its neutralization (Abrol *et al.*, 1975).

B. Fertilizer Management

The crops receiving sodic water irrigations are adversely affected unless additional nitrogen is applied to compensate for losses due to denitrification (Dargan and Chillar, 1978; Van Hoorn, 1958). Increased nitrogen application compensate the yield reduction due to increasing levels of exchangeable sodium percentage (Abrol, 1968). At CSSRI Karnal, studies on the responses of applied nitrogen to rice and wheat in sodic soils indicated that these crops respond to higher level of nitrogen than those grown on nonsodic soils (Annual Reports 1970-1980). Similar results were reported by Obrejanu and Sandhu (1971) in sugarbeet and Yadav and Dhankhar (!983) in wheat. These experiences suggest that the crops irrigated with sodic waters be fertilised at 15 to 20% higher rates recommended for normal soils. Increased uptake of Ca and Mg and decreased uptake of sodium resulting from application of additional nitrogen were responsible for the greater response to N at high exchangeable sodium levels. Rao and Batra (1983) reported that as the N loss is minimum at 20 ppm N (40 kg N/ha), N should be applied to wheat in 3-4 equal splits of 120 kg N/ha to optimise the N deficiency in sodic soils. They observed that complementry anions of the ammonical fertilizers such as in ammonium sulphate, ammonium nitrate and ammonium carbonate as also in diammonium phsophate did not influence the N loss through volatilization. Application of nitrogenous fertilizer such as ammonium sulphate and urea resulted in similar increase in yield of rice and wheat (Singh, 1987). It is observed that in tropical sodic soils, slow release and other nitrogenous fertilisers do not respond to nitrification inhibitors (Rao and Batra, 1984). Placement of nitrogenous fertilizer 6-7 cm deep for wheat can reduce the volatilisation losses by 90 per cent (Rao and Batra, 1983).

Results of a long term fertility trial in a sandy loam soil indicated that rice-wheat sequence on sodic soils need no P and K fertilisation in the initial few years of reclaimation (Chhabra, 1985). Although the application of only P fertiliser has not shown any crop response, in combination with Zn fertilisation it increased the crop yield due to the interactive P-Zn effect after 2 years of cropping in sodic soils. Cop response to applied K fertilisers has not generally been observed in sodic soils (Dargan and Chhilar, 1973; Chhabra, 1985) due to predominance of micaceous minerals in the soils of many arid and semi arid regions (Sindhu and Gilkes, 1977;

Kapoor *et al.*, 1981).

To make up Zn deficiency in sodic soils, application of 10 kg $ZnSO_4$/ha is found effective in rice (Singh *et al.*, 1982). Marked response (25-100%) to Zn application were observed in sorghum-wheat, pearlmillet -wheat, lady's finger-berseem and paddy-wheat rotation under low SAR (8.0) and high RSC (7.8 me L^{-1}) sodic water (EC 1 dSm^{-1}) used for 4 years on loamy sand soil at Gurgaon (Manchanda and associates, unpublished) and in pearlmillet at Bawal (Yadav, 1991). Large gain in yield on Zn fertilisation of sodic soil has been attributed to its improvement in tolerance of crops to sodic environments (Shukla and Parsad, 1974). The difference between the effective and ineffective amounts of Zn seems to vary widely, depending on the chemical nature and pH of the sodic soils (Singh *et al.*, 1983). Since retrogradation of precipatates limit adequate supply of Zn to crops at high pH (Singh and Abrol, 1986), regular application of Zn in smaller doses seems more appropriate. However, addition of iron salt to make up the deficiency was generally unfruitful unless it was accompained by changes brought about in oxidation status of soil by continuous submergence and addition of organic matter. (Katyal and Sharma, 1980; Swarup, 1980).

C. Rain Water Harvesting

The effect of sodic water on soil properties is modified considerably according to rainfall pattern of the area. Chauhan et al., (1988) found that rain water created condition conducive for the amelioration of the soil. They argued that after rains, there was decrease in soil properties viz; EC, ESP, and pH and increase in hydraulic conductivity, possibily due to leaching process (Table 28). Similar results have been reported by Manchanda and Bhandari (1976).

In a lysimetric study, Suneja (1985) observed that 200 mm simulated rain cycle decreased ESP of 0-30 cm soil by 16.7, 13 and 9 percent, respectively in the sandy, sandy loam and clay soil. Lal and Singh (1973) observed that with the advent of rain (391.4 mm), the respective ESP in the 4 layers of profile (0- 22.5, 22.5-45.0, 45.0-67.5 and 67.5-90 cm) decreased by 33, 41, 44 and 38 per cent. Summer ploughing is very useful practice to increase infiltration, to check run-off and to conserve soil moisture in the problematic sandy desert soils (Malik *et al.*, 1989). The increase was much more when infiltrating water was of high RSC (Table 29). They further argued that ploughing increased the value of the infiltration characterstics, A (hydraulic conductivity of the transmission zone), S (sorptivity) and I_{80} (cumulative infiltration at 80 minutes).

Table 28 : Seasonal variations in physicochemical properties of the soil.

Levels of RSC (me L^{-1})	Initial (Oct. 81)	After March 82	After rains Sept. 82	After harvest March.83	After rains Sept. 83	After harvest March.84
			ECe (dSm^{-1})			
5	1.7	4.4	1.2	4.1	1.6	4.0
10	1.7	4.7	1.3	4.3	1.6	4.0
15	1.7	4.3	1.3	4.3	1.7	4.0
20	1.7	4.0	1.2	3.8	1.5	3.2
			pH			
5	8.6	8.6	8.3	8.6	8.5	8.6
10	8.6	8.8	8.6	8.7	8.6	8.7
15	8.6	8.8	8.7	8.9	8.8	8.9
20	8.6	9.0	8.8	8.9	8.8	9.0
			ESP			
5	9.2	15.4	11.1	18.6	12.7	18.5
10	9.2	19.1	12.7	21.3	14.0	22.3
15	9.2	28.0	13.4	28.7	14.8	29.6
20	9.2	33.8	14.1	34.2	15.2	35.5
		Hydraulic	conductivity	($cm\ h^{-1}$)		
5	1.5	1.6	1.6	1.5	1.6	1.5
10	1.5	1.5	1.6	1.5	1.6	1.5
15	1.5	1.3	1.6	1.2	1.4	1.2
20	1.5	1.2	1.4	1.1	1.4	1.1

Table 29 : Percent increase of infiltration characterstics on ploughing under low (L) and high (H) RSC infiltration waters

RSC	A	S	I_{80}
L	344	73	228
H	485	333	377

Run off percentage from ploughed plots were much lower than those from ridged plots. Disking the ploughed soil without leveling and ridging increased the run-off markedly. On the plowed undisked plots the run off was reduced due to the roughness of soil surface, yet one pass with a heavy disk reduced surface roughness to an effective extent equal to that of the ridged plots (Johnson *et al.*, 1979).

D. Selection of Crops and Varieties

The growth, yield and sometimes quality of most crop species is affected with the use of sodic waters. There exists a wide genetic variation in crops and their varieties in relation to sodicity (Kumar *et al.*, 1983; Kumar and Abrol, 1986; Sharma and Manchanda, 1989). Yadav *et al.* (1991b) tested five major *rabi* crops under sodic water condition and concluded that their relative tolerance was rated as barley > wheat > raya > field pea > chickpea. Similarly, Pearson and Bernstein (1958) studied the effect of exchangeable sodium on six crops and found that the relative tolerance of the crops studied was found as tall wheat > grass > barley > rice > tall fescue > oats. They further argued that the level of injury varied with soil texture. At equivalent ESP values, the crops (wheat and barley) were more seriously affected on clay than on loams, whereas for oats the opposite relationship was observed.

Therefore, a farmer in selecting a crop species to be grown in such area, has to be very careful to get satisfactory returns. After the selection of crop his next job is to select a variety which is relatively more tolerant to sodicity and also productive. A list of crops that can be grown over a sodicity range is given in Table 30.

Table 30 : Relative tolerance of some crops and grasses to exchangeable sodium

Tolerant	Semi tolerant	Sensitive
Karnal grass	Wheat	Cowpea
Rhodes grass	Barley	Chickpea
Para grass	Oats	Groundnut
Bermuda grass	Mustard	Lentil
	Senji	Black gram
Rice	Pearlmillet	Maize
Dhaincha	Cotton	Cotton (at germination)
Sugarbeet	Berseem	Green gram
	Sugarcane	Peas.

Source : Abrol. 1990

Crop yields are seriously affected if the ESP is more than about 55,35 and 10 in respect of tolerant, semi-tolerant and sensitive crops, respectively. The grasses listed are highly tolerant and some, like Karnal grass, will grow even in soils of ESP 80 to 90 (Abrol, 1990).

E. Plant Population

The phase of germination and seedling growth is a critical one, as the ability of a crop to germinate and establish the seedlings frequently becomes a limitng factor in crop production. But it is not always true that if a crop is tolerant at germination, it will remain tolerant throughout its life cycle. Some of the crop species which are tolerant at later stages of growth are quite sensitive during germination and vice- versa (Bernstein and Hayward, 1958).

Proper crop stand is the most important factor for obtaining higher yield. Plant population in the soils irrigated with the sodic waters can be obtained by increasing the seed rate or the number of transplant in rice. Gupta *et al.* (1988) reported that survival of tree seedlings is affected with the rise in sodicity and the deleterious effect was maximum in loam and minimum in clay soil. Increased seed rate makes up the reduction in germination and mortality of seedling during early growth stages (Dargan *et al.*, 1974).

Germination of crop plant is adversely affected by increasing levels of ESP (Pearson and Bernstein, 1958) and the severity of injury in seedlings varied among the several species, oats being affected at 36%, barley at 48% and wheat at 54% exhcnangeable sodium. Reports indicate that germination is adversely affected with sodic water irrigation in pulses (Yadav *et al.*, 1990b), oilseeds (Chhabra *et al.*, 1979) and medicinal plants (Sharma *et al.*, 1988). The germination of seed is markedly improved by gypsum application. Thus, proper plant population in soils irrigated with sodic waters can be obtained by increasing seed rate and with the use of amendment depending upon the severity of the problem.

Surface crusts have also been observed to cast deterimental effect on seedling emergence and soils irrigated with sodic waters are more prone to crusting. The differential emergence behaviour of test crop under similar set of conditions may be attributed to crop factors like time taken for seed germination, epicotyl size and emergence force. In addition, soil drying and cracking patterns have also been found to affect the crop performance (Arndt, 1965) which in turn is affected by soil ESP (Pinauli, 1982). Pinauli (1982) found no significant effect of ESP on crust strength upto ESP 30. Pinauli and Abrol (1986) studied the effectiveness of polyacrylamide (PAM) in improving seedling emergence of some selected *kharif* and *rabi* crops on soidc soils. They reported that on an average, the emergence under PAM was about 13% higher than control. They observed that the test crops differed significantly in their emergence behaviour and reported cowpea among kharif crop and gram and pea among *rabi* crops as the least adversely affected crops. They observed that mustard, wheat and bajra seedling were unable to over come crust problem due to their fragility. In maize, emergence was poor due to delay in germination process though it had higher emergence strength as evidenced by Prihar and Agarwal (1975).

F. Planting Technique

Soil irrigated with sodic water have very low infiltration rate thus water remains standing in the field for long. Due to poor soil aeration, the crop root system is disturbed. It has been observed that in such soils, highest root and sucrose yields are obtained in the methods where sugarbeet is planted on both sides of flat ridge (Singhania *et al.*, 1989). Similar results were obtained by Garg and Srivastava (1985). They attributed this to the proper aeration and tendency of salts to move towards the centre of flat ridge. It is suggested that the planting should be done by seed because the roots of transplanted seedlings were fangy and ill-shaped (Panje and Gill, 1969; Garg and Srivastava, 1985). It has been observed that under sodicity, raising of pearlmillet was more beneficial when transplanted (Manchanda and Sharma, 1984) on ridges (Yadav *et al.*, 1989, 1991) as compared to direct seeding (Table 31).

Table 31 : Effect of different sowing methods on grain yield (q ha^{-1}) of pearlmillet under sodic water condition

Sowing method	Direct seeding	Transplanting	Mean
On flat surface	15.5	19.9	17.7
On ridges	21.0	23.3	22.6
In furrows	13.6	12.2	12.9

Source : Yadav et al. (1989, 1991).

G. Crop Rotation

It is observed that in wheat- pearlmillet/sorghum rotation, addition of gypsum significantly enhanced the grain and stover yield of pearlmillet and sorghum whereas, it could not increase wheat yield in wheat-fallow rotation (Sharma and Manchanda, 1989). They reported that when soil ESP was less than 16, the yield of pearlmillet and sorghum remained unaffected but it decreased considerably (30-40%) when soil ESP in the rhizosphere (0-30 cm) was 27. Bajwa and Josan (1989) reporated that decline in crop yields with increasing sodicity was more pronounced under rice-wheat than millet- wheat rotation. The results of the long term experiment (Table 32) conducted on loamy sand soil in south western Haryana on the use of sodic waters (RSC 2.8, 11 and 16 me L^{-1}) indicated that lower yield was obtained in wheat-pearlmillet than in wheat-fallow rotation (Dhankhar *et al.*, 1990). They attributed it to development of higher pH and ESP in the field.

Table 32 : Effect of different RSC on grain yield of wheat and pearlmillet in two rotations.

RSC levels me L^{-1}	Wheat grain yield (g ha^{-1})				Pearlmillet grain yield (q ha^{-1})	
	1984–85		1985–86		1984	1985
	Fallow wheat	Pearlmillet wheat	Fallow wheat	Pearlmillet wheat	Fallow wheat	Pearlmillet wheat
2.8	42.2	37.5	52.2	50.4	29.7	17.1
11.0	33.3	28.3	47.9	43.0	24.6	8.1
16.0	19.4	16.4	38.9	35.8	20.9	3.2

H. Biological Methods

Although several plants have been used for alleviating the sodium hazards of irrgation water and soil sodicity itself, alfalfa has received general acceptance (Kanwar, 1962). It has a great tolerance to sodicity and water stagnation. Its extensive root system open up the soil to allow better aeration and leaching. Rapid decomposition of roots and acidic nature of sap help in neutralising the harmful effect of sodicity for successful growing of subsequent crops. Plants like *Chenopodium* spp. absorb large amount of salts and thus help in scavanging Na from the field if uprooted and thrown away. Similarly, sugarbeet tops accumulates 5 times more sodium than roots (Garg, 1987) and help in reducing the field sodicity, if tops are either thrown or fed to animals. Growing of salt tolerant forage grasses particularly Karnal grass lowered the pH of sodic soils (Kumar and Abrol, 1984). The rate of CO_2 release from the roots of growing plants is an indication of soil reclaimation process. (Goertzen and Bewers, 1958). It is observed that the CO_2 evolution from pot soil increased 2 times with submergence alone and 6 times when combined with grass growth (Grewal and Abrol, 1989).

IX. CONCLUSIONS

On the basis of available literature and foregoing discussion on various aspects of the problems it may be concluded that continuous use of sodic water inflict a harmful effect on physical and chemical properties of soil and affect the nutrient availability in soil. Plant growth is severely restricted due to impaired soil health. The reduction in crop yields is brought about either by nutritional imbalances or osmotic effect or by direct ion injury. Satisfactory crop yields can be harvested when these waters are judiciously used with suitable amendments and fertilizer doses in proper crop rotation depending on soil type under a specific climatic situation. Selection of tolerant crops and their varieties suited to sodic water irrigation when grown with suitable agrotechniques help a big way in harnessing these waters.

REFERENCES

Abrol. I.P. (1968). Trans. 9th Int. Congr. Soil Sci. 2 : 585.

Abrol. I.P. (1990). Soil fertility and fertilizer use. Vol. IV IFFCO :. P183.

Abrol. I.P. and Acharya. C.L. (1975). Proc. 2nd World Congr. Int. Water Resource Assoc. New Delhi, India. Dec., 1975 (1) : 335.

Abrol. I.P. and Bhumbla. D.R. (1979). Soil Sci. 127 : 79.

Abrol. I.P., Dahiya. I.S. and Bhumbla. D.R. (1975). Soil Sci. 120 : 30.

Abrol. I.P., Khosla. B.K. and Bhumbla. D.R. (1968). Geoderma 2 : 33.

Acharya. C.L. and Abrol. I.P. (1978). Soil Sci. 125 : 310.

Aggarwal. R.R. and Mehrotra. C.L. (1953). Soil Survey and Soil Work in U.P. Vol. 3. 337 P. Supdtt.. Printing & Stationary U.P. Allahabad.

Agarwala. S.C., Sharma. C.P., Sinha. B.K. and Malhotra, N.K. (!964). J. Indian Soc. Soil Sci. 12 : 343.

Agarwal, R.R., Singh. M. and Pal. C. (1950). Bull. Deptt. Agric. U.P. 104.

Alam, M., Prasad, C.R. and Singh, K.D.N. (1986). J. Indian Soc. Soil Sci. 34 : 855.

Annual Reports (1970-1980). CSSRI. Karnal.

Ansari, R. (1972). Pak. J. Bot. 4 : 55.

Arndt, W. (1965). Aust. J. Soil Sci. 3 : 45.

Ayers, R.S. and Westcot, D.W. (1985). F.A.O. Irrigatin and Drainage Pap. 29.

Bajwa, M.S., Hira, G.S. and Singh, N.T. (1983). Irrig. Sci. 4 : 191-199.

Bajwa, M.S. and Josan, A.S. (1989). Agricultural Water Management. 16 : 53-61.

Bajwa, M.S., Josan, A.S., Hira, G.S. and Singh, N.T. (1986). Irrig. Sci. 7 : 27.

Bajwa, M.S., Singh, N.T., Randhawa, N.S. and Brar, S.P.S. (1975). J. Res. Punjab Agric. Univ., Ludhiana, India, 12 : 117.

Bernstein, L. and Hayward, H.E. (1958). Ann. Rev. Plant Physiol.9: 25.

Bhumbla, D.R. and Abrol, I.P. (1972). Indian Fmg. 21 (4) : 15.

Bhumbla, D.R. and Dhingra, D.R. (1964). J. Indian Soc. Soil Sci. 12 : 255-60.

Biswas, T.D., Samra, J.S. and Dutt, O. (1979). Bull. Indian Soc. Soil Sci. 12 : 23.

Bower, C.A., Ogata, G. and Tucker, J.M. (1968). Soil Sci. 106 : 29.

Chang, C.W. (1961). Soil Sci. 91 : 29.

Chhabra, R. (1985). Agron. J. 77 : 699.

Chhabra, R., Singh, S.B. and Abrol, I.P. (1979). Soil Sci. 127 : 242.

Chhabra, R., Singh, A. and Abrol, I.P. (1980). Soil Sci. Soc. Am. J. 44 : 33.

Chauhan, R.P.S., Bhudayal and Chauhan, C.P.S. (1988). J. Indian Soc. Soil Sci. 36 : 590.

Chauhan, R.P.S., Chauhan, C.P.S. and Dixit, H.C. (1989). J. Indian Soc. Soil Sci. 37 : 431.

Chawla, K.L. and Chhabra, R. (1991). J. Indian Soc. Soil Sci. 39 (1) : 40.

Cope, J.T. (Jr.), Bradfield, R. and Peech, M. (1953). Soil Sci. 76 : 65.

Dargan, K.S., Abrol, I.P. and Bhumbla, D.R. (1974). Agron. J. 66 : 279.

Dargan, K.S. and Chhilar, R.K. (1973). Indian Fmg. 22 (12) : 13.

Dargan, K.S. and Chhilar, R.K. (1978). Indian J. Agron. 23 : 14.

Deo, R. (1964). M. Sc. thesis, Punjab Agric. Univ. Hisar, India.

Dhakhar, O.P., Yadav, H.D. and Yadav, O.P. (1990). National Symposium on 'Water Resource Conservation, Recyling and Reuse'. Nagpur, India P. 57-59.

Dhir, R.P., Sharma, B.K. and Singh, N. (1980). Int. Symp. Salt Affected Soil CSSRI, Karnal, India, 369.

Dravid, M.S. and Goswami, N.N. (1986). J. Indian Soc. Soil Sci. 34 : 87.

Eaton, F.M. (1950). Soil Sci. 69 : 123.

Ehrler, W. (1960). Bot. Gaz. (Chicago) 122 : 102.

Frankenberger, W.T. (Jr.), and Bingham, F.T. (1982). Soil Sci. Soc. Am. J. 46 : 1173.

Garg, V.K. (1988). J. Indian Soc. Soil Sci. 35 : 343.

Garg, V.K. and Srivastava, S.C. (1985). J. Indian Soc. Soil Sci. 33 : 884.

Genkel, P.A. and Bakanova, L.V. (1974). Fiziol. Rast. 21 : 812.

Goertzen, J.O. and Bowers, C.A. (!958). Proc. Soil Soc. Am. 22 : 36.

Grewal, S.S. and Abrol, I.P. (1988). Indian J. agric. Sci. 58 (4) : 278.

Grewal, S.S. and Abrol, I.P. (1989). J. Indian Soc. Soil Sci. 37 : 371.

Gupta, R.K., Bhumbla, D.R. and Abrol, I.P. (1984). Soil Sci. 137 : 245.

Gupta, G.N., Prasad, K.G., Mohan, S., Subramaniam, V. and Manivachakam, P. (1988). J. Indian Soc. Soil Sci. 36 : 537.

Gupta, I.C. (1979). Use of saline water in agriculure in arid and semi arid zones of India, Oxford and IBH Publishing Co., New Delhi.

Gupta, I.C. and Abichandani, C.T. (1970). J. Indian Soc. Soil Sci. 18 : 429.

Gupta, R.K. and Ranade, D.H. (1987). J. Indian Soc. Soil Sci. 35 : 474.

Gupta, R.K. and Verma, S.K. (1983). Curr. Agric. 7 : 114.

Gupta, S.K. and Sharma, S.K. (1990). Irrig. Sci. 11 : 173.

Haunsenbuiller, R.L., Haque, M.A. and Wahhab, A. (1960) Soil Sci 90 : 357.

Hira, G.S. and Singh, N.T. (1980). Soil Sci. Soc. Am. J. 44 : 930.

Johnson, C.B.: Mannering, J.V. and Moldenhaner, W.C. (1979). Soil Sci. Soc. Am. J. 43 : 772.

Jones, L.H.P. and Leeper, G.W. (1951). Plant and Soil. 3 : 154.

Kanwar, B.S. (1964). Ph. D. thesis, Punjab Agric. Uni. Hisar, India.

Kanwar, B.S. and Kanwar, J.S. (1971). Indian J. agric. Sci. 41 : 154.

Kanwar, J.S. (1962). Symp. Salinity and Alkalinity, IARI, New Delhi.

Kanwar, J.S. and Kanwar, B.S. (1968). Trans. 9th. Int. Cong. Soil Sci. Adelaid Australia. 1 : 391.

Kanwar, J.S., Sehgal, J.L. and Bhumbla, D.R. (1963). J. Indian Soc. Soil Sci. 11 : 39-51.

Kanwar, J.S. and Singh, S.S. (1961). Soil Sci. 92 : 207.

Kapoor, B.S., Singh, H.P., Goswami, S.C., Abrol, I.P., Bhargava, G.P. and Pal, D.K. (1981). J. Indian Soc. Soil Sci. 28 : 486.

Katyal, J.C. and Sharma, B.D. (1980). Plant and Soil. 55 : 105.

Kavimardan, S.K., Budhe, N.N. and Ballal (1964). J. Indian Soc. Soil Sci. 12 : 281.

Kelly, W.P. (1962). Soil Sci. 94 : 1.

Kemper, W.D., Olsen, J. and De Moory, C.J. (1975). Soil Sci. Soc. Am. Proc. 39 : 458.

Keren, R. and O'Connor, G.A. (1982). J. Soil Sci. Soc. Am. 46 : 726.

Keren, R. and Singer, M.J. (1989). Soil Sci. Soc. Am. Proc. 53 : 349.

Kesvan, S.P. and Gupta, R.K. (1986). J. Indian Soc. Soil Sci. 34 : 442.

Kumar, A. and Abrol, I.P. (1984). Indian J. agric. Sci., 54 : 189.

Kumar, A. and Abrol, I.P. (1986). Tech. Bulletin 11. CSSRI. Karnal. 95 pp.

Kumar, D., Singh, M.P. and Buttar, B.S. (1983). Curr. Agric. 7 : 97-113.

Kurian, T. (1976). Z. Pflanzenphysiol. 79 : 373.

Lagerwerff, J.V. and Eagle, H.E. (1961). Plant Physiol. 36 : 77.

Lal, P., Malik, G.C. and Singh, K.S. (1980). Ann. Arid Zone. 19 (4) : 359.

Lapina, L.P. (1966). Fiziol. Rast. 13 : 1029.

Larson, W.E. and Pierre, W.H. (1953). Soil Sci. 76 : 51.

Leeper, G.W. (1947). Soil Sci. 63 : 79.

Lessani, H. and Andreopulas-Renand, U. (1970). Acad. Sci. Paris (D). 269-577.

Levit, J. (1972). Salt and Ion Stress. Acad. Press, New York. p489.

Malik, R.S., Johorar, B.S., Singh, B. and Butter, B.S. (1986). Int. Symp. Water Manage. Arid. Semi-Arid Zone. 33-43.

Malik, R.S., Yadav, H.D. and Yadav, O.P (1989). International Symposium on "Managing Sandy soils. CAZRI. Jodhpur. India. p. 96.

Manchanda, H.R. (1976). Quality of ground waters in Haryana. Haryana Agric. Univ., Hisar. P. 160.

Manchanda, H.R. and Bhandari, D.K. (1976). J. Indian Soc. Soil Sci. 24 : 432.

Manchanda, H.R., Garg, R.N., Sharma, S.K. and Singh, J.P. (1985). J. Indian Soc. Soil Sci. 33 : 876.

Manchanda, H.R., Gupta, I.C. and Jain, B.L. (1989). Int. Symp. Managing Sandy Soils. Central Arid Zone Res. Inst. Jodhpur, India, February 6-10, 1989. P. 362.

Manchanda, H.R., Sharma, S.K. and Singh, J.P. (1985). J.Indian Soc. Soil Sci. 33 : 366.

Manchanda, H.R., Sharma, S.K., Yadav, S., Buttar, B.S. and Suneja, B.K. (1985). Indian Fmg. 35 : 9.

Manchanda, H.R., Verma, S.L. and Khanna, S.S. (1982). J. Indian Soc. Soil Sci. 30 (3) : 353.

Martin, J.P. and Bingham, F.T. (1954). Soil Sci. 78 : 349.

Martin, J.P., Jones, W.W. and Ervin. (1959). Agron. J. 51 : 418- 421.

Mehta, K.K. and Abrol, I.P. (1976). Indian Fmg. 26 (1) : 21.

Mitra, S.P. and Shanker, H. (1957). Soil Sci. 83 : 471.

Nimbalkar, J.D. and Joshi, C.V. (1975). J. Biol. Sic. 18 : 55.

Obrejanu, G. and Sandhu, G.A. (1971). 1. Szabolcs (ed). European solonetz Soils and their Reclaimation. Akademiai Kiado. Budapest. P. 99.

Page, E.R. (1962). Plant and soil. 16 : 247.

Painuli, D.K. (1982). Ph. D. Thesis Kurukshetra Univ. Kurukshetra. India.

Painuli, D.K. and Abrol, I.P. (1986). J. Indian Soc. Soil Sci. 34 : 179.

Pal, D.K. and Mondal, R.C. (1980). J. Indian Soc. Soil Sci. 28 : 347.

Pal, R. and Poonia, S.R. (1979). J. Indian Soc. Soil Sci. 27 (1) : 5.

Paliwal, K.V. and Gandhi, A.P. (1976). Soil Sci. 122 : 85.

Paliwal, K.V. and Gandhi, A.P. (1977). Fertilizer Technology. 14 (1&2) : 32.

Panje, R.R. and Gill, P.S. (1969). Indian Sugar 19(2) : 169.

Paricha, P.C., G.J. and Sahoo, P. (1975). Acta. Bot. Sinica. 12 : 64.

Pasricha, N.S. and Randhawa, N.S. (1971). Int. Symp. on Soil Fertility Evaluation. pp. 1017.

Patel, K.P. and Singh, B. (1991). J. Indian Soc. Soil Sci. 39 (1) : 154.

Partric, W.H.Jr. and Wyatt, R. (1964). Soil Sci. Soc. Amer. Proc. 28 : 647.

Pearson, G.A. and Bernstein, L. (1958). Soil Sci. 86 : 254.

Ponnamperuma, F.N. (1972). Adv. Agron. 24 : 29.

Poonia, S.R., Mehta, S.C. and Pal, R. (1984). J. Indian Soc. Soil Sci. 32 : 663.

Poonia, S.R. and Pal, R. (1979). Agric. Water Management. 2 : 163.

Poonia, S.R., Pal, R., Nath, J. and Singh, M. (1978). In : Land and water Management in the Indian basin. 1 : 230.

Pratt, P.F. and Thorne, D.W. (1948). Soil Sci. Soc. Am., Proc. 12 : 213-217.

Prihar, S.S. and Agerwal, G.C. (1975). Soil Sci. 120 : 200.

Puntamkar, S.S., Mehta, P.C. and Seth, S.P. (1972). J. Indian Soc. Soil Sci. 20 (3) : 281.

Raghav, R. (1991). M. Sc. thesis. Haryana Agric. Univ. Hisar, India.

Rai, Y., Singh, D., Sahi, B.P., Singh, K.D.N. and Prasad, M. (1980). Indian Sugar. 29 : 639.

Rao, D.L.N. and Batra, L. (1983). Plant and Soil. 70 : 219.

Rao, D.L.N. and Batra, L. (1984). Int. Rice Res. Newsletter. 9 : 26.

Rhoades, J.D. (1972). Soil Sci. 113 : 277.

Richards, L.A. (1954). Diagnosis and Improvement of saline and Alkali soils. USDA, Handbook No. 60. P. 160.

Saha, A.K. (1976). Ph. D. thesis. Haryana Agric. Univ. Hisar, India.

Sahrawat, K.L. (1980). Fertilizer News. 25 (8) : 12.

Sandhu, S.S., Acharya, C.L. and Abrol, I.P. (1980). Int. Symp. on "Salt affected Soils". Feb. 18-21, 1980, CSSRI. Karnal : 183-194.

Sandhu, S.S., Acharya, C.L. and Abrol, I.P. (1981). J. Indian Soc. Soil Sci. 29 : 148.

Sawhney, J.S. and Baddesha, H.S. (1989). J. Indian Soc. Soil Sci. 37 : 418.

Sengupta, M.B. and Maji, A.K. (1977). Fertilizer Technology. 14 (1&2) : 38.

Shah, R.K. and Palkhiwala, A.B. (1979). Curr. Agric. 3 : 53.

Shahi, M.N., Maskina, M.S. and Gill, P.S. (1978). Plant and Soil. 49 : 437.

Sharma, B.M. and Yadav, J.S.P. (1986). J. Indian Soc. Soil Sci. 34 : 849.

Sharma, B.M. and Yadav, J.S.P. (1988). J. Indian Soc. Soil Sci. 138 : 141.

Sharma, M.L. (1971). Aust. J. Soil Res. 9 : 73.

Sharma, M.L. (1973). Agron. J. 65 : 982.

Sharma, O.P., Trivedi, K.C., Gupta, R.K., Paradkar, V.K. and Gangrade.

S.K. (1988). J. Indian Soc. Soil Sci. 36 : 774.

Sharma, S.K. and Manchanda, H.R. (1989). J. Indian Soc. Soil Sci. 37 : 135.

Shukla, U.C. and Mukhi, A.K. (1980). Proc. Int. Symp. Salt Affected Soils. Karnal, India. P. 362.

Sindhu, P.S. and Gilkes, R.J. (1977). J. Soil Sci. Soc. Am. 41 : 1194.

Singh, B. and Bajwa, M.S. (1987). J. Indian Soc. Soil Sci. 35 : 358.

Singh, D., Tripathi, R.K., Vilay, P.K., Mediratta, J.C. and Bansal, K.N. (1980). J. Indian Soc. Soil Sci. 28 : 213.

Singh, G. (1987). Symposium on Alternate Farming System. Indian Society of Agronomy. N. Delhi. P. 120.

Singh, K.D.N., Prasad, C.R. and Singh, Y.P. (1986). J. Indian Soc. Soil Sci. 34 : 152.

Singh, K.S., Lal, P. and Lal, P. (1967). J. Indian Soc. Soil Sci. 15 : 35.

Singh, K.S. and Sharma, R.P. (1970). J. Indian Soc. Soil Sci. 18 : 345.

Singh, M., Poonia, S.R. and Pal, R. (1988). J. Indian Soc. Soil Sci. 36 : 330.

Singh, M.V. and Abrol, I.P. (1980). Plant and Soil. 94 : 445.

Singh, M.V., Chhabra, R. and Abrol, I.P. (1982). Annual Report CSSRI, Karnal.

Singh, M.V., Chhabra, R. and Abrol, I.P. (1983). Soil Sci. 136 : 359.

Singh, S.B. and Abrol, I.P. (1985a). J. Indian Soc. Soil Sci. 33 : 358.

Singh, S.B. and Abrol, I.P. (1985b). Plant and Soil. 84 : 123.

Singh, S.B. and Abrol, I.P. (1988). J. Indian Soc. Soil Sci. 36 : 316.

Singh, S.B., Chhabra, R. and Abrol, I.P. (1979). Agron. J. 71 : 767.

Singh, S.B., Chhabra, R. and Abrol, I.P. (1980). Indian J. agric. Sci. 50 : 852.

Singh, S.B., Chhabra, R. and Abrol, I.P. (1981). Indian J. agric. Sci. 51 : 885-891.

Singhania, R.A., Mishra, N.P. and Sharma, R.K. (1989). J. Indian Soc. Soil Sci. 37 : 421.

Solov'ev, V.A. (1969). Fiziol. Rast. 16 : 498.

Strogonov, B.P. (1964). Acad. Sci. USSR. Davey and Co. New York.

Suarez. (1984). Soil Sci. Soc. Am. J. 45 : 469.

Suneja, B.K. (1985). Ph. D. thesis. Haryana Agric. Univ. Hisar, India.

Swarup, A. (1980). J. Indian Soc. Soil Sci. 28 : 532.

Swarup, A. (1985). J. Indian Soc. Soil Sci. 33 : 938.

Swarup, A. (1986). J. Indian Soc. Soil Sci. 34 : 198.

Thein, S.J. (1976). Soil Sci. Soc. Am. J. 40 : 105.

Thomas, G.W. and Yaron, B. (1968). Soil Sci. 106 (3) : 213.

Troug, E., Berger, K.C. and Attre, O.J. (1953). Soil Sci. 76 : 41.

Uppal, H.L. (1955). Indian J. agric. Sci. 25 : 211.

Van Hoorn, J.W. (1958). Netherlands J. Agric. Sci. 6 : 1.

Venkatakrishnan, S. and Abrol, I.P. (1981). J. Indian Soc. Soil Sci. 29 : 526.

Verma, S.K. and Gupta, R.K. (1985). J. Indian Soc. Soil Sci. 33 : 465.

Verma, S.K., Gupta, R.K. and Dubey, S.K. (1985). J. Indian Soc. Soil Sci. 33 : 948.

Wilcox, L.V., Blair, G.V. and Bower, C.A. (1954). Soil Sci. 77 : 259.

Yadav. H.D. (1989. 90) Seasonal Reports. HAU. Regional Research Station. Bawal, India.

Yadav. H.D. and Dhankhar. O.P. (1989). Seasonal Report. HAU. Regional Research Station. Bawal. India.

Yadav. H.D. and Dhankhar. O.P. (1987). Seasonal Report. HAU. Rgional Research Station. Bawal. India.

Yadav. H.D.. Dhankahr. O.P. and Phogat. V.K. (1985). Curr. Agric. 9 : 55.

Yadav. H.D.. Kumar. V.. Singh. S. and yadav. O.P. (1991). H.A.U.. J. Res. In press.

Yadav. H.D.. Kumar. V.. Malik. R.S. and Yadav. A. (1991). Ann. Arid Zone. 30 : 373

Yadav. H.D., Sultan Singh and Yadav. O.P. (1990). International Symposium "On National Resource Management for a sustainable agriculture" 284 (Feb. 6-10; 1990).

Yadav. H.D., Yadav. O.P.. Dhankhar. O.P. nd Kumar. V. (1990). Ann. Arid Zone. 29 (4) : 275.

Yadav. J.S.P. and Aggarwal. R.R. (1961). J. Indian Soc. Soil Sci. 9 : 151.

Yadav. J.S.P. and Mehta. K.K. (1980). Ann. Arid Zone. 19 (3) : 359.

Yadav. O.P. (1991). Personal communication. Soil Scientist. HAU. NARP. Gurgaon. India.

Dryland Resources & Technology, 8 (1994)
Editors : L.L. Somani, K.N. Bansal & R.A. Sharma

MANAGEMENT OPTIONS FOR VERTISOLS AND ASSOCIATED SOILS OF CENTRAL AND PENINSULAR INDIA

R. A. Sharma
AICRP (Dryland Agriculture)
JNKVV, College of Agriculture, Indore - 452 001.

I. INTRODUCTION

Out of 275 million hectares of black clay soils (Vertisols and associated vertic soils) occurring throughout the world, 76.4 million hectares lie in India (Murthy, 1981). These soils, however, are concentrated in central and peninsular regions between 8°45' to 26° 0'N latitude and 66° O' to 83° 45'E longitude covering an area of about 73 million hectares. This constitutes 22.2 per cent of total geographical area of the country. Distribution of these soils in different states of India is depicted in Table 1. These soils are bestowed with high moisture retention capacity ranging from 350 to 410 mm per meter (Table 2) and favourable water release characteristics due to high clay content (more than 55 per cent) dominated by montomorillonite type of clay mineral. They are characterized by low infiltration rate varying from 0.2 to 29 mm per hour (Table 3), high degree of plasticity and stickiness, low organic matter, high cation exchange capacity, calcareous nature and alkaline reaction. Dominance of smectite group of clay minerals leads to a considerable swelling and shrinkage on wetting in rainy season and drying during post rainy season, poor internal drainage and narrow workable moisture range. Although the black soils have enormous potential productivity both under irrigated and unirrigated conditions, large hectrage remains underutilized owing to inherent management problems during rainy season apart from limited supply of irrigation water during post riany season period and nutrient limitations. An attempt has been made in this paper to review the constraints to management and productivity of black clay soils and available technological options pertaining to conservation and management of these soils.

II. CONSTRAINTS TO MANAGEMENT AND PRODUCTIVITY

Dominance of montmorillonite clay in black soils leads to high degree of swelling on wetting and shrinkage on desiccation due to which arise the problems of poor infiltration, high runoff, soil erosion and nutrient losses during rainy season and enhanced evaporative water losses due to development of shrinkage cracks during prolonged dry spell in rainy season and post rainy season. Areas having dependable rainfall (more than 750 mm annual rainfall) coupled with high intensity and erratic temporal and spatial distribution frequently results in waterlogging on flat lands and runoff and severe soil erosion on slopy lands.

Table 1 : Distribution of vertisols and associated soils in India (Murthy, 1981)

Name of state	Total area (m ha)	Percent of gross area in India	Per cent of total geographical area in India
Maharashtra	29.9	35.5	7.9
Madhya Pradesh	16.7	23.0	5.1
Gujarat	8.2	11.9	2.6
Andhra Pradesh	7.2	10.0	2.2
Karnataka	6.9	9.4	2.1
Tamilnadu	3.2	4.2	1.0
Rajasthan	2.3	3.0	0.7
Orissa	1.3	2.0	0.4
Bihar	0.7	1.0	0.2
Uttar Pradesh	Negligible	Negligible	Negligible
Total	76.4	–	22.2

Table 2 : Water storage in some black soils of central India at sowing and harvesting of unirrigated wheat crop.

Location	Profile depth (cm)	Water content (mm) at sowing	at harvest	Available water (mm)	Source
Jabalpur*	100	350	207	143	Jaggi (1975)
	150	550	320	230	
	180	650	404	246	
Indore	100	409	270	239	Sharma (1976)
	150	600	412	188	
	180	720	492	228	

The major factors contributing to severity of soil erosion are land slope, runoff, and traditional systems of cultivation particularly the practice of rainy season fallowing. The runoff data presented in Table 4 emphasize the strong need for scientific water management alongwith soil erosion control measures. In view of poor infiltrability of these soils, widely occurring undulating topography, monsoonic nature of rainfall and fallowing of land in many areas for obvious reasons, runoff is inevitable in the black soils regions. Tables 4 and 5 depict the long term data of rainfall, runoff and rainfall intensity and duration recorded at Indore, the western part of the state of Madhya Pradesh. Runoff water, a precious natural resource, not only erodes the soils, washes away plant nutrients and limits the productivity in these areas but also creates the flood problems and damages the crops in lower reaches.

Table 3 : Terminal infiltration rates of black clay soils of central and peninsular India.

Location	Infiltration rate (mm/hour)	Source
Indore, Madhya Pradesh	12.6	Gupta and Verma (1983)
Jabalpur, Madhya Pradesh	4.7	Gupta et al (1976)
Hyderabad, ICRISAT, A. P.	0.21	Krantz et al (1978)
Rahuri, Maharashtra	6.4	Shinde et al (1982)
Parbhani Series, Maharashtra	7.0	Ambegaonkar et al (1984)
Satna, Madhya Pradesh	4.0	Gupta et al (1984)
Bijapur, Karnataka	2.3	Anonymous (1987a)
Kovilpatti, Tamilnadu	29.0	Anonymous (1987b)

Table 4 : Rainfall and runoff from medium and deep black soils at Indore, Madhya Pradesh.

Medium black soil				Deep black soil			
Years	Raifall (mm)	Runoff (mm)	Runoff (%)	Year	Rainfall (mm)	Runoff. (mm)	Runoff (%)
1971	975	523	53.6	1979	675	140	20.7
1972	539	184	34.1	1980	973	165	16.9
1973	1889	936	49.5	1981	840	94	11.2
1974	868	182	21.0	1982	629	4	0.7
1975	1376	398	28.9	1983	987	40	4.0
1976	1241	524	42.2	1984	879	206	24.3
1977	1131	417	36.9	1985	636	17	2.7
1978	735	389	49.5	1986	928	201	21.7
1979	675	270	40.0	1987	762	48	6.3
Mean	1053	425	40.3	Mean	812	102	12.6

Table 5 : Maximum rainfall intensities (mm/hour) for varying duration of storms recroded once in different years (Shrivastava, 1987).

Years	Time in minutes					
	5	10	15	30	45	60
1977	96	–	80	68	39	39
1978	120	81	72	62	44	36
1979	180	144	112	72	40	32
1980	300	150	140	110	100	82
1981	144	102	100	98	–	50
1982	240	132	124	96	70	68
1983	96	–	80	50	48	45
1984	120	78	76	50	39	38
Expected once in 10 years	157	145	136	121	80	58

Soil erosion is the serious problem in black soils with land slope greater than 0.8 per cent while on flat lands sheet erosion as well as waterloggig are the serious problems. Data presented in Table 6 indicate the losses of soil and nutrients from black clay soils at Indore. These data emphasize the necessity of soil and water conservation measures.

Table 6 : Soil and nutrient losses from black clay soils at Indore, Madhya Pradesh (Anonymous, 1991).

Slope (%)	Runoff (mm)	Soil Loss (kg/ha)	Nutrient loss (kg/ha)	
			N	S
1.0	69	613	10.67	2.84
1.5	85	918	15.31	4.37
2.0	116	986	23.85	3.00

The major constraints to high productivity of black clay soils arise due to their peculiar physico-chemical characteristics. These constraints can be grouped into three categories namely; physical, chemical and biological constraints. Physical constraints are; (i) low infiltration and poor internal drainage, (ii) narrow workable moisture range of these soils, (iii) poor crop stand, (iv) evaporation from soil surface and shrinkage cracks, (v) runoff and associated soil erosion and nutrients losses, (vi) salinity and alkalinity hazards, (vii) non- availability of drainage design criteria etc.

Chemical constraints are : (i) high values of soil reaction i.e.

high soil pH, which adversely affects the availability of plant nutrients such as P, Fe, Mn and Zn, (ii) calcareous nature of soils which enhances volatilization losses of N following surface application of ammonium and ammonium forming fertilizers (Goswami and Sahrawat, 1982), (iii) soils are low in organic carbon and loss of N from urea and ammonical fertilizers further aggravate the situation, (iv) wide spread deficiency of P in Indian soils (Tandon, 1987). High P fixing capacity of black clay soils limits the desired response of applied fertilizer P, (v) very little attention paid to K management, (vi) emerging S deficiency in many areas, and (vii) deficiency of certain micronutrients particularly of zinc and iron.

As far as biological constraints are concerned, potential for maximizing biological N fixation is underutilized. Further, the importance of biological N fixation has been suppressed by fertilizer N in intensive and high input agriculture (App *et al.*, 1980). App *et al.* (1980) have listed four biological constraints for underutilization of biological N fixation. These are; (i) fertilizer N has been less expensive and is far easier to manage than biological N_2 fixation, (ii) yield of grain legumes are generally low and they usually have not responded to high level of management, (iii) grain legumes are generally grown on marginal lands where appropriate technology for legume based farming systems is not available, and (iv) there is general lack of understanding about the potential for symbiotic nitrogen fixation.

III. SOIL CONSERVATION PRACTICES

Some land treatments and water disposal systems have been found to control soil erosion from black clay soils. Broadbed and furrows, graded furrows, broad based bunds, diversion bunds, loose boulder structures and bench terracing are the commonly recommended treatments. Contour bunding has been used as a soil conservation measure in many places, but such conservation structures have been reported to adversely effect crop yields (Kampen, 1974). Some promising soil conservation measures are discussed below.

A. Reclamation and Control of Gullies

Masonary structures are costly and they do not work in swell-shrink type of soils as their life is very short in these soils for obvious reasons. Gabion structures to reclaim gullies and stabilise waterways, and grassing of waterways along natural depressions have been successfully used on a field scale in deep black soils of western Madhya Pradesh (Verma and Raje, 1981). A 'Gabion' is a semi-flexible structure of loose boulders packed into prefebricated galvanised iron wire netting and serves as a drop structure. Drop structure of various sizes and slopes can be made by placing a number of boxes (Gabion) together at the top of one another as per requirement. Gabions have been found to be effective than masonary structures of the same design as they are porous, flexible and allow the runoff water to pass through while retaining the silt upstream.

B. Bunds

Bunds of various kinds for different purposes and situations have been found effective to check erosion losses from black clay soils (Verma, 1981). Graded bunds are constructed to control soil erosion on lands having slope in the range of 1 to 6 per cent. These are constructed across the slope on a grade of 0.1 to 0.2 per cent spaced at suitable vertical intervals, so as to divide the total catchment area into small parts and runoff water from each part is carried away along these bunds without causing any erosion. The optimum vertical interval (spacing) for graded bunds for land slope upto 1.5%, 2%, 3%, 4%, 5% and 6% has been reported as 1.3 m, 1.5 m, 1.7 m, 1.8 m, 2.0 m, and 2.2 m respectively by Gadkary (1966).

Diversion of runoff water comming off a hillock into cropped fields saves the land from degradation. This can be achieved by constructing a bund with a shaped drainage channel along it on a grade of 0.2 to 0.3 per cent. The cross section of the bund will depend upon the rate of runoff and catchment area contributing runoff to be diverted. The base width of 1.0, 1.5, 2.3, and 3.5 meters for 20, 30, 40 and 50 hectares catchment area, respectively has been suggested by Gadkary (1966). The optimum side slope of a bund is considered to be 1.5 to 2 : 1. If the catchment area includes a bare hillock, the base width has to be increased by about 50% and pitching with boulders at the ends may also be necessary to take care of greater amount of runoff.

C. Bench Terracing

Bench terracing coupled with plantation of trees and grasses on lands having more than 6 per cent slope is an effective measure of erosion control. Bench terraces cut down the velocity of flowing water, retain soil and thus retards the soil erosion by allowing safe disposal of runoff water to a drainage channel.

D. Grassed Waterways

Grassed waterways are required to be constructed at suitable sites to lead water diverted by storm drains or water diversion bunds and graded bunds to natural stream. The cross section of a waterway would depend upon the rate of runoff, catchment area and the value of permissible velocity of flowing water which depends upon the nature of bed. The permissible velocity of flowing water for a bare sandy soil, bare clay soil, grassed soil and *Murrum* (weathered rock) and hard rock are 0.5, 1.0, 1.4 and 1.7 m/seconds, respectively (Verma, 1981). If the bed of waterway is too deep, it gets eroded when the velocity of water exceeds these limits and in such cases the bed of waterway should be stabilised by constructing appropriate Gabion structures at proper points.

E. Stabilisation of Washes

Washes formed in the cultivated fields may develop into gullies if not controlled in initial stages. Stabilisation of washes can be achieved by leaving that strip of land uncultivated over which water flows so that grasses may get established in due course of time. For quick results, suitable grasses may be planted initially. If wash is in advanced stage of erosion and getting deeper and wider every year, it may be necessary to provide drop structures. For a shallow wash (depth upto 50 cm) a rectangular weir type drop structure of loose boulders at appropriate site with reverse filter (consisting of pebbles, gravel and sand layers) upstream from the structure will be sufficient. For deeper washes (depth more than 50 cm), appropriate Gabion structures are necessary.

F. Provision of Drainage Between Waterways

Safe disposal of runoff water from areas between waterways is essential. For this, open drainage channels on a grade are to be constructed at suitable intervals from ridge to grassed waterway. The cross section of drainage channels will depend upon the rate of runoff and the areas of its catchment. Depending upon the difference in the bed level of the grassed waterway and those of

Table 8 : Effect of weed management on pearlmillet – legume intercropping system

Treatments	Yield (q/ha)				Weed dry weight (q/ha)	
	1985–86		1986–87			
	Mirzapur farm		B.H.U. farm		1985–86	1986–87
	Pearl-millet	Black-gram	Pigeon-pea	Pearl-millet		
1. Weedy check	2.66	0.89	4.75	6.51	177	37
2. Farmer's method	3.34	1.06	5.01	10.12	69	9
3. Interculture twice	3.71	1.28	5.32	9.54	84	31
4. T3 + intra row handweeding	3.73	2.13	6.51	10.42	67	15
5. Fluchloralin 1.0 kg a.i./ha	1.44	1.90	5.63	8.24	125	36
6. T5 + one interculture at 30 DAS	1.48	2.23	8.42	9.26	115	27
7. Prometryn 1.0 kg a.i./ha	3.01	2.35	8.15	8.24	55	28
8. T7 + one Int. at 30 DAS	6.05	2.52	9.93	10.56	37	24
C.D. 5%	1.39	1.33	3.52	2.57	71	19

Table 9 : Effect of tillage and seeding device on the crop establishment of barley

Treatment	Yield (q/ha)		
	88–89	89–90	Mean
Tillage			
Deep ploughing (Twice)	4.11	8.56	6.34
Mould board ploughing (Twice)	3.85	9.02	6.44
Cultivator (Twice)	4.47	9.92	7.20
Disking (Twice)	5.43	10.95	8.19
C.D. 5%	0.28	0.93	
Seeding devices			
Behind plough	4.09	8.47	6.28
Fepso plough	4.42	9.44	6.93
Malviya seeddrill	4.88	10.91	7.90
C.D. 5%	0.54	0.87	

V. INPUT MANAGEMENT

Comprehensive field studies at Dryland Research Centre, B.H.U., Varanasi, clearly indicate that improved technology (integration of fertilizer application and plant protection measures against aphids) has provided five times more yield of mustard as compared to traditional technology (Table 10). The magnitude of yield advantage under various inputs has been varying. Experiences from these studies make it obvious that two production factors, i.e. fertilizer use and aphid control measure, assume a striking importance for improving the productivity of mustard crop. Their effect was more productive when applied in combination than in isolation.

Table 10 : Performance of mustard under integrated input management

Treatment	Varuna Yield (q/ha)			Pusa Barani (q/ha)		
	89-90	90-91	Mean	89-90	90-91	Mean
1. Variety only	2.11	0.68	2.59	3.06	0.79	0.74
2. Management only (Plant protection for aphids)	2.54	3.70	4.41	4.28	4.23	3.97
3. Fertilizer only	10.74	2.11	10.79	10.84	2.38	2.25
4. Fertilizer + management	11.74	2.93	11.95	8.46	11.95	8.20
Mean	6.78	3.61		7.59	3.97	

Biological factors in conjunction with socio-economic factors determine the complexities of cropping systems in Drylands. The environmental constraints determine the productivity of these systems and research activities constitute an attempt to improve the performance of the system. An integrated approach to the problem helps to synthesize information in ways that permit application to system. The informations set out in the paper gives an impact of integrated management approach towards optimising basic resources in conjunction with management practices. Lot more is required to be done in future on system research with an approach towards integration.

VI. RESEARCH GAPS AND FUTURE RESEARCH NEEDS

Compost, FYM and green leaf manuring are age old systems but are brideled with the problem of logistics. The review of research work set out in the foregoing pages reveals that many research gaps are yet to be filled up in evolving out nitrogen management practices for multiple cropping systems particularly under dryland conditions. Evidences of direct transfer of nitrogen from the legume components are yet to be emerge clearly.

Uncertainity in rainfall both in terms of quantity and distribution makes fertiliser use in drylands risky. Integrated nitrogen management system involving alternate sources of nitrogen should be harnessed, so that reasonable yields from drlylands can be obtained even without the use of chemical fertilisers. This can certainly help the cause of drylands but for increased and sustained productivity there will be no option left out to advocating integrated management practices. Results have been comming in this respect but the issue needs to be settled by initiating qualitative research involving nutrient recovery and soil physical properties are lacking.

VII. SUMMARY AND CONCLUSIONS

The dryland of eastern Uttar Pradesh is beset with intractable problems related to climate, soils, crops and socioeconomic aspects. Agriculture is risk prone, farmers lack resources, drought power is limited, tools are limited and inefficient and land productivity is low and unstable in such areas. An integrated approach for conserving soil and water resources has been generated to a good extent for improving productivity. Now important aspect is its adoption by the farmers which, in turn, contributes to boost total agricultural production and uplifting the quality of the farmers.

REFERENCES

Gigou. J. 1986. Fertiliser use on sorghum in West Africa. Pages 379-404. In Proceedings of the second regional seminar on crop production techniques and fertiliser management in rainfed agriculture in Southern Asia. New Delhi: World Phosphate Institute.

Leggett, G.E. 1959. Relationships between wheat yield. avaiable moisture, and available nitrogen in eastern Washington dryland area. Bulletin 609. Washington Agricultural Experiment Stations: Pullman Washington State University.

Lotheretz, W. 1989. Open questions in sustainable agriculture. Am. J. Alternative Agriculture. 3 : 174-181.

Narkhede, P.L.; Umroru, N.K. and Ghugare, R.V. 1984. Organic recycling of Leucaena leucocephala for winter sorghum cultivated on drylands. Indian J. Agril. Sci., 5 : 444-446.

Parthoharajono, S. Hordjono, A. 1986. Food and State crop production problems and fertiliser management in rainfed areas in Indonesia. Pages 129-160. In proceedings of the second regional seminar on crop production techniques and fertiliser management in rainfed agriculture in southern Asia, New Delhi, World Phosphate Institute.

Prihar, S.S. and Arora, V.K. 1980. Crop response to mulching with crop residues. Bulletin Deptt. of Soils Ludhiana Punjab Agril. Univ.

Randhawa, N.S. and Singh, R.P. 1983. Feriliser management in rainfed areas-available technologies and future needs. Fertilizer News 28(9): 17-32.

Rao, A.C.S. and Das, S.K. 1982. Soil fertility management and fertiliser use in dryland crops. Page 120-139. In a decade of dryland agricultural research in India, 167-180.

Singh, R.P., Singh, J.P., Singh, Y., Singh, A.K. and Singh, R.A. 1989. Weed management in rainfed rice-lentil crop sequence. IRRN Vol. 14 (2), p. 39-40.

Unger, P.W. 1978. Straw mulch rate effect on soil water storage and sorghum yield. Soil Science Society of American Journal, 42 : 486-491.

Unger, P.W. and Parker, J.J.Jr. 1976. Evaporation reduction from soil with wheat, sorghum and cotton residues. Soil science Soc. of America Journal, 40 : 938-942.

Venkateswarlu, J. 1983. Nutrient supply from organic sources. p. 54-60. In proceedings National Seminar on dryland agricultural research in India, thrust in the eighties.

Dryland Resources & Technology, 8 (1994)
Editors : L.L. Somani, K.N. Bansal & R.A. Sharma

AMELIORATION OF ALKALI SOILS THROUGH FLY ASH

*L.L. Somani, K.L. Totawat & M.C. Mundra**
Rajasthan College of Agriculture,
Udaipur - 313 001

I. INTRODUCTION

The feature of industrial progress of the developed nations of the world has been a simultaneous increase in their energy use pattern and they are trying to keep pace with their growing commercial energy needs. Of all the commercial energy sources, coal has been largely used in preference to others. A vast quantity of coal is being burnt in thermal power plants for the purpose of power generation in addition to its use in other industries and

* CCS-HAU, Hisar

railways. The utilization of coal in power generation results in production of huge quantities of fly ash, the particulate coal waste product of thermal power plants. Furthermore, use of low grade coal and inferior model of ESP (Electro Static Precipitator) a greater amount of fly ash is evolved creating land degradation problems and may cause potential environmental hazards in spreading diseases and leading of unwanted chemicals into the eco-system. The residual ash has to be removed and at present no use has been found for it. Among the several amendments, gypsum and pyrite are suitable for reclamation of alkali soils, but it is gradually becoming costly. There is, however, good possibility of using fly ash as a soil amendment in the reclamation of alkali soils as it is a potent source of calcium, sulphur and magnesium (Mahalingam, 1973).

II. COMPOSITION OF FLY ASH

Since fly ash is partly or wholly a burnt material it has some of the properties of coal as well as burnt ash. A detailed chemical composition of fly ash is given in Table 1. Accordingly fly ash is acidic in nature and contains a high percentage of sulphur, calcium and magnesium with small percentage of nitrogen, phosphorus and potassium. These constituents may prove good for reclaiming sodic soils as well as may prove good for improving the fertility of soils in due course of time (Mahalingam, 1973; Plank and Martenes, 1973; Singh and Singh, 1986).

Table 1 : Chemical analysis of fly ash

Particulars	Values
pH	4.1
EC (dS/m)	0.514
Available nitrogen (%)	0.003
Total nitrogen (%)	0.011
Available phosphorus (%)	0.006
Available potassium (%)	0.008
Organic carbon (%)	0.210
Sulphur (%)	5.040
Ca as CaO (%)	1.050
Mg as MgO (%)	1.030

Source : Singh and Singh (1986)

III. AMELIORATIVE MECHANISM

A. Increased Availability of Nutrients

Fly ash supplies considerable amount of nutrients in forms readily available to plants. Better supply of nutrients creats most favourable condition for vigorous and luxuriant plant growth which ultimately improve growth and yield (Singh and Singh, 1986). The ameliorative influence of fly ash could also be attributed to enhanced availability of nutrient by bringing down the soil pH (Mahalingam, 1973).

B. Replacement of Exchangeable Sodium by Calcium

The rich content of calcium in fly ash undergo dissolution in soil solution and as a consequence of valence dilution rule, it replaces sodium from the exchange complex of alkali soils.

C. Improvement in Soil Physical Properties

Rajamanner *et al.* (1971) reported increased organic carbon content as well as increased bulk density of soil following incorporation of fly ash. Witkovszky (1966) stated that fly ash when mixed with N P K and lime proved as an effective substitute for dung in increasing the humus level of soil and improving the structure and moisture content of soil, thereby, counteracting the adverse effect of high alkalinity. Herke (1959) stated that fly ash prevents premature wilting of rice especially in alkali soil and also it showed a prolonged residual action in the soil. The soluble calcium of the fly ash provides congenial atmosphere for the flocculation of highly dispersed alkali soil particles. The organic matter content of this ash provides much needed protective action to stabilise the physical environment improved by the calcium.

Application of fly ash to light textured soils will increase their water holding capacity as are being rich in finer fraction and have very high (72.0 to 80.2%) water holding capacity (Aitken *et al.*, 1984, Maiti *et al.*, 1990).

D. Neutralisation of Alkalinity

Being acidic, fly ash directly brings down pH by neutralising carbonates, bicarbonate and hydroxyl ions. In addition to it, sulphur which is an essential component of fly ash, produces sulphuric acid upon microbial oxidation. This sulphuric acid can directly neutralise the alkali. Besides, H_2SO_4 so produced may react with native $CaCO_3$ to bring calcium in soluble form to replace sodium from the exchange complex.

Results of several experiments conducted in India and abroad have clearly shown that incorporation of fly ash caused marked reduction in the pH of the treated soils (Herke, 1959; Adams, *et al.*, 1973; Barber, 1979; Mahalingam, 1973) (Table-2).

IV. CROP RESPONSE TO FLY ASH

Since fly ash contains a high percentage of Ca, Mg and S with small percentage of N,P and K, it possesses the potential of being a good amendment for reclaiming alkali soils besides improving the fertility of the treated soil (Plank and Martenes, 1973). All these constiuents will be of great help in improving the ultimate crop production potential of the soils. Bodganovic *et al.* (1964) observed significant improvements in the yield of maize and tobacco by application of lignite at 40 and 60 g/pot respectively. An increase in the yield to the extent of 25 per cent over mineral fertilizers has also been reported for maize and wheat. A similar increase in the yield of hay crop *Puccinellialimosa* has also been reported (Herke, 1959).

Table 2 : Effect of fly ash levels on soil pH and EC (dS/m)

Treatment	1980-81		
	25 days after application of fly ash	After harvest of rice crop	After harvest of wheat crop
	pH		
Control	7.87	7.82	7.80
10%	7.67	7.60	7.54
20%	7.29	7.18	7.17
30%	7.28	7.14	7.13
SEm ±	0.02	0.01	0.01
CD at 5%	0.05	0.04	0.04
	EC dS/m		
Control	0.347	0.339	0.338
10%	0.854	0.861	0.855
20%	0.985	0.974	0.969
30%	1.025	0.990	0.983
SEm ±	0.0045	0.002	0.001
CD at 5%	0.013	0.007	0.003

Source : Singh and Singh (1986)